MATH
Strategies You Can Count On

MATH
Strategies You Can Count On

Tools & Activities to Build Math Appreciation, Understanding & Skills

by Char Forsten

Crystal Springs
BOOKS
A division of SDE Staff Development for Educators
Peterborough, New Hampshire

Published by Crystal Springs Books
A division of Staff Development for Educators (SDE)
10 Sharon Road, PO Box 500
Peterborough, NH 03458
1-800-321-0401
www.crystalsprings.com
www.sde.com
© 2005 Crystal Springs Books
Published 2005

Printed in the United States of America
09 08 07 06 05   1 2 3 4 5

ISBN: 1-884548-71-7

Library of Congress Cataloging-in-Publication Data

Forsten, Char, 1948-
  Math strategies you can count on : tools & activities to build math
appreciation, understanding & skills / by Char Forsten.
       p. cm.
  Includes bibliographical references and index.
  ISBN 1-884548-71-7
  1. Mathematics—Study and teaching. 2. Teaching—Aids and devices. I. Title.
  QA16.F67 2005
  372.7--dc22
                        2004027171

Editor: Sharon Smith
Art Director, Designer, and Production Coordinator: Soosen Dunholter
Photography: Soosen Dunholter

D edicated to Lorraine Walker, who shares my enthusiasm in the quest to teach and learn math from many angles, and whose vision inspired me to write this book.

Contents

Acknowledgments

First and foremost, I am indebted to my remarkable editor, Sharon Smith. Her extraordinary talent, encouragement, and patience are what made this book possible.

I continue to be in awe of Soosen Dunholter's creative, artistic work, and I am so grateful that she designed this book.

I wish to thank Deb Fredericks, whose exceptional organizational skills kept this book and all it entails on track.

I also owe a very special thank you to Catherine Kuhns for reading my book for mathematical content. I respect and admire her expertise in math and value her comments and suggestions.

Introduction

Imagine trying to go through a day without using mathematics. You couldn't check the time, or keep it for that matter. Schedules would be a matter of guesswork. Forget money. And don't bother checking on today's temperatures, stock prices, sports scores, or department-store sales. If you did shop, at best you could ask only for *some* milk, meat, and vegetables. I'm not sure a store owner would even allow you to leave with groceries because—remember—you wouldn't be able to use money if you tried to get through a day without math.

Of course, you'd also need to eliminate all polygons and other shapes from your wardrobe and environment. Bridges would tumble without all of those magnificent triangles supporting them. And, oh yes, there are the patterns that surround us, such as the sunrise, sunset, and tides—the predictability of those precise mathematical phenomena that we *count* on every day would disappear.

Clearly, the point here is that our world is richly and magnificently mathematical. We could not possibly live without math. Yet how many of our students appreciate this fact? How many students make the connections between their everyday lives and the math concepts and skills they learn in school—or between their surroundings and those concepts and skills? As teachers, we can maximize opportunities for students to make these meaningful connections by providing some simple yet powerful tools and ways of looking at things.

It's important to find ways to help students discover the math around them. Not only will the mathematical concepts make more sense to the learner, but he'll also gain a stronger understanding and appreciation of math. Motivating students to appreciate and value mathematics is a major goal that I address in this book, especially in the chapter "Warm-Ups & Cool-Downs."

Another important goal is to improve students' mathematical communication skills. By mathematical communication, I'm referring not only to reading, creating, and interpreting charts and graphs. I believe we also need to provide frequent opportunities for students to revisit and use mathematical vocabulary. Too many upper-elementary students still suggest that you "plus" or "minus" instead of using the terms "add" and "subtract." The activities in the "Warm-Ups & Cool-Downs" chapter provide ways for students to use specific math vocabulary, and to use it not just on the day it is taught, but on a regular basis.

So I start by encouraging students to recognize, appreciate, and value the math that surrounds them; and by building their mathematical vocabulary. As those goals are met, I find that student self-confidence increases as well. All of this contributes to reaching my predominant goal: building each student's sense of number.

I want to take the "numb" out of number and put wonder in its place. When a student has a strong number sense, he has a solid foundation for effective problem solving. And having the skills to solve mathematical problems in the real world is what number sense is all about.

I truly believe that students need different pathways to learn and learn well. In my teaching, I discovered that many of the ways the textbook presented math concepts and skills weren't necessarily resulting in meaningful student learning. This led me to seek alternatives that would resonate with my students—the kinds of tools and strategies I've included in this book.

Every activity presented here is designed to engage all your students by offering a variety of pathways to success. These alternative pathways are critically important to our teaching. We cannot say, "I taught it, so he got it."

We also need to build different types of ongoing assessments into our daily teaching. As a teacher, I was always trying to determine each student's level of learning. I would try to establish whether the student was at the recognition stage, at the reproduction or modeling stage, or at the independent production stage of learning a particular skill or concept.

Too often, we provide direct instruction, then guided practice, and finally independent practice. At that point, we assume a student has internalized the learning, and we move on.

For example, let's say I'm teaching the addition of unlike fractions. First, I provide direct instruction. Next, over a period of a few days, the students practice working on examples under my guidance, with a focus on the procedural steps involved in adding unlike fractions. Finally, students work independently through more examples in class or for a homework assignment. I might even give a quiz or test, and they might score well. The critical question is whether they can independently add unlike fractions weeks or months from now.

Have you ever had a student come running into your classroom and beg, "Please give me the math test quick, before I forget"? This may be the same student who moved quickly through the recognition, reproduction/guided practice, and independent production stages. She appeared to understand in class and may have done her assignments correctly, but she may have stored the content only in short-term memory. It's possible that student may not even recognize unlike fractions two weeks later.

That's why we must determine what essential learning needs to take place, at what level it should be learned, how it's best taught, and how we can ensure that students actually register it in long-term memory. For most students, this means we must begin with meaningful, concrete experiences to develop understanding at the conceptual level. Then we need to build on this foundation by teaching procedures to solve specific math problems. One size does not fit all, and as teachers, we need options.

This book is a compilation of different pathways to learning numerous elementary mathematical concepts and skills. I hope the strategies, activities, and tools that have helped many of my students learn and love mathematics will also translate into the wonder of number for your students.

Suggested Group Sizes

The activities in this book are very adaptable to a range of group sizes. These are my suggestions for the group size(s) for which each activity is appropriate. These are intended strictly as broad guidelines; only you know what's best for your class.

	Whole	Small	Pair	Individual
1. Warm-Ups & Cool-Downs				
Have You Seen Any Good Math Lately?	●	●		
Have You Read Any Good Math Lately?	●	●		
Have You Heard Any Good Math Lately?	●	●		
Have You Eaten Any Good Math Lately?	●	●		
Geometry Scavenger Hunt	●			
Overhead Operations	●	●		
Renaming the Date	●			
Daily Number Review	●	●		
Number Stars	●			
Guess My Number	●	●		
Thinkersize #1: Picture Puzzles	●	●		
Thinkersize #2: Flexible Fred	●	●		
Thinkersize #3: Hink Pinks	●	●		
2. Focusing & Response Tools				
Page Protectors	●	●	●	●
Focus Frames	●	●	●	●
Envelope Windows			●	●
Highlighting Tape	●	●	●	●
Response Cards	●	●		
Dry-Erase Tools	●	●	●	●

	Whole	Small	Pair	Individual
3. Learning Math Through Patterns				
Sum Patterns			●	●
Triangular Flash Cards		●	●	●
Adding & Subtracting Two-Digit Numbers		●	●	●
Multiple Patterns & Finding Common Multiples	●	●	●	●
Finding Averages Through Patterns	●	●	●	●
Focusing on the Facts	●	●	●	●
Skip Counting & Multiples	●	●	●	●
Another Way to Find Common Multiples	●	●	●	●
Equivalent Fractions	●	●	●	●
Adding & Subtracting Unlike Fractions	●	●	●	●
Divisibility Rules	●	●	●	●
4. Mastering Math Facts All Year Long				
Leveled Card Packets	●	●	●	●
Captive Dice	●	●	●	●
Building a Tetrahedron & a Hexahedron	●	●		●
The Facts of Life	●	●	●	
First with the Facts	●	●	●	
Facts on the Brain	●	●	●	
Object Math	●	●	●	●
Smath	●	●		
5. "Discovering" Math Skills & Concepts				
Place Value Participation	●	●		
Subtraction Action	●	●		●
Finding Square Numbers & Their Roots	●	●	●	●
Partner Factors	●	●	●	●
Guesstimate/Bestimate	●			
Putting Statistics on the Line	●			
Measurement—What Shape Are You?		●	●	
What's Your Circumference?		●	●	

Chapter 1

WARM-UPS & COOL-DOWNS

Warm-Ups and Cool-Downs are five-minute mental math activities intended to "grease the neurons" for students' mathematical thinking. The Warm-Ups and Cool-Downs build in fun, meaningful review. They provide opportunities for students to practice, thus reinforcing mathematical skills and concepts, and building computational fluency. They also boost students' higher-level thinking skills by teaching mental math strategies. Teacher Cathy Kuhns has observed that "students cannot talk math until they've heard their teacher talk math." I agree. So these activities also provide numerous opportunities for you to "talk the talk" of math with your students.

I like to think of Warm-Ups and Cool-Downs as being very much like an exercise routine. You warm up by spending five minutes on one of these activities at the beginning of a math class. The lesson is the workout, and then you pull things back together for students with the Cool-Down.

You can use Warm-Ups and Cool-Downs all year in your class, following the same basic activity structure while increasing the complexity level of the skills the class is learning. A real advantage to these opening and closing activities is that they can address a wide range of math concepts and skills. They also allow you to open math class on a motivational note. Asking students, "Have you seen any good math

lately?" is a wonderful way to stir students' creative thinking and help them connect what they're learning to real life.

In this chapter, you'll find numerous options for Warm-Ups and Cool-Downs. My experience has been that these activities tend to work best when you choose just one Warm-Up and one Cool-Down each day. That's enough to engage students and help them prepare for and review the day's new mathematical challenges.

Have You Seen Any Good Math Lately?

Build observational skills; reinforce students' understanding of patterns, geometry terms, and other math concepts; and help students connect math to the real world with this creative, educational activity.

Preparation

Cover the "root" label at the base of the reproducible and make a transparency of the page to use as graph paper in your modeling.

Collect in a big box all the appropriate printed material (such as the types of items in the "Materials" list) you can find. Anything is fair game as long as it includes photographs or other artwork that can be cut up.

Make a heading for the class bulletin board that asks the question, "Have you seen any good math lately?"

MATERIALS

- Reproducible #1
- Photographs
- Postcards
- Magazines
- Photos and art prints from old calendars
- Other printed materials
- Digital camera (optional)
- Computer and printer (optional)

Directions

This is an interactive bulletin board. Start by asking students, "Have you seen any good math lately?" Then, to model how they can answer that question, hold up this book and ask them to look at the window in the photograph on page 112. (If you prefer, substitute a photo of your own.) Point out that, as in the copy shown at right, you can see the multiplication array of 2 X 3; there are two rows of windowpanes, with three in each row.

Challenge students to find other math in the photograph. Someone might offer that he sees parallel lines. Another student might say she sees rectangles, right angles, and fractions. Encourage a variety of answers so everyone understands that there is no single correct answer and everyone is invited to participate.

This is a good time to share with students two other books that include excellent models of math photographs. Arlene Alda's *1 2 3 What Do You See?* (1998, Tricycle Press) and Tana Hoban's *Shapes, Shapes, Shapes* (1986, HarperCollins) capture students' imaginations and can become wonderful springboards for their own personal versions of math in the everyday world.

Another source of endless fascination for children and adults alike is the work of M.C. Escher. Perspective, proportion, angles, and shapes are all mathematical connections to the world of art. Escher's work captures all of these—and at the same time captures students' imaginations with his intriguing transformations and tessellations.

Once students understand how this activity works, invite them to dive into the box of pictures and print materials, and "find the math" to post on the bulletin board. Make it clear that they need to be able to explain their answers. During your Warm-Up or Cool-Down session, ask one student at a time to share what she's found. You might also ask students to capture their observations in journals, with or without photographs, as a highly effective way to practice communicating mathematically.

Ask students to bring other examples in from home. Old calendars often have great photos or artwork for this (imagine the potential in a calendar with photos of quilting blocks!). Magazine ads can be terrific sources, too, and so can postcards. Once students understand what to look for, they'll come up with all sorts of great examples. Post a sampling on the bulletin board, and during your next Warm-Up or Cool-Down, ask other students what they can find in these pictures.

I've found that this strategy increases students' creative, open-ended thinking dramatically. More important, all students participate and learn from each other. When one student looks at the photograph of the window and points out symmetry and right angles in the picture, other students learn from that observation.

This type of collaboration exemplifies what Harry and Rosemary Wong tell us in their book, *The First Days of School* (1998, Harry Wong Publications): "Learning is an individual activity, but not a solitary one. It is more effective when it takes place within a supportive community of learners."

Once students are accustomed to this activity, turn it into tear-free homework, either as an "always can do" or as an assigned task. Some students will become really excited about looking for wonderful patterns. I remember one insightful student who shared a photo of a road with a solid median strip, broken white lines, and telephone poles and lines along its edge. (A similar photo appears at right; see page 113 for a larger version you can share with the class.) The student observed that the median strip is the middle line in the road, just as the median number is the middle number in a set of values.

Other students found more math; they pointed out that the broken white lines represented line segments and that the wires ran parallel to the road and perpendicular to the telephone poles.

That photo became part of my instructional materials. I used it for years because it was also a great visual to help students understand the concept of a line in geometry. By definition a line has no beginning and no ending point. That holds no meaning for some students. It's an abstract concept that is difficult to conceptualize. But the outstretched, straight road is a representation. You can't see the beginning or the end of the road, so a photo like this helps students to understand.

It gets even better! With digital cameras and inkjet printers, students can create their own books filled with photographs that capture math in everyday objects and places. Assign small groups of students to walk around the school, the playground, or the town, cameras in hand. Math is everywhere.

During another session of "Have You Seen Any Good Math Lately?" a girl in my class brought in a photograph that looked something like the picture on page 114. She recognized that this was "too many to count"—a perfect example of when you should estimate. We looked for other math in the photo, too. This is the type of activity that lets you push the students' thinking. You can say, "Now come on, I'm seeing something else in here that you know. A pumpkin resembles what kind of solid? That's right! A sphere."

Eventually, you can demonstrate a hands-on estimation strategy with the pumpkins. Just put your graph paper transparency over the pumpkins and say, "Well, could we estimate by figuring out about how many pumpkins are in one square, and then determining how many squares have pumpkins?" Explain that once we have this information, we can multiply the number of squares that show pumpkins times the approximate number of pumpkins in each square and arrive at an estimated total.

Have You Read Any Good Math Lately?

This simple activity teaches students to use real-world documents and information to look for math and to solve problems. It also raises their math awareness levels as they start looking for representations in environmental print.

Directions

When you first ask students if they've read any good math lately, they'll probably look at you cross-eyed. That's when you bring in your newspaper and model for them how to "read math." *USA Today* is great for this. Students can find math in the date, the price, the sports section, the TV section, the stock reports, and especially in the full-page, color weather map of the United States.

MATERIALS

- Copies of *USA Today*
- Copies of local newspaper

Whatever students need to do well, they need to do frequently. We can develop fluency in map reading by providing regular opportunities for students to use the maps. Have students find the city closest to them on the weather map, then note that day's high and low temperatures in a daily journal. Soon they'll begin to find their location on a map automatically.

Next, still referencing the weather map, ask students to locate those cities with the highest and lowest temperatures in the nation that day. Then ask them to calculate the difference in those two temperatures.

Of course, you can find all kinds of numbers in the sports section. Ask students to use the problem-solving strategy of making an organized list to show the different ways a football score might have been achieved. Ask, "If the score was 21 to 15, what are the possible combinations of touchdowns, extra points, and field goals scored?"

Don't stop there! In the TV section, ask students to determine how long different programs last. Have them examine the classifieds or the real estate section of a local newspaper and note differences in prices. Or have students choose virtually any part of the paper and practice working with appropriate levels of place value; ask them to find numbers and to highlight the hundreds in one color, the tens in a second color, and the units in a third.

Now move beyond the newspaper to other environmental print: road maps, signs, posters, and labels. Have students note and report distances in miles or kilometers, the shapes of road signs, numbers on posters, phone numbers on billboards, symmetry in labels, and patterns they see on all kinds of environmental print. Reading and interpreting nutrition and ingredient labels on packages will also encourage students to use factual math information in meaningful discussions about healthy diets.

Be sure, too, to have students take another look at any books they're currently reading. Again pose the question, "Have you read any good math lately?" Encourage them to notice page numbers, rhyming patterns, charts and graphs, shapes, and of course any actual math concepts.

You can also use children's literature to further reinforce math concepts. Introduce your students to *The Brainy Day Math Book Series*, published by Scholastic. These wonderful stories connect several math concepts to situations and plots that make the math memorable and understandable in a way a planned skills lesson cannot. Books in the series include *The Greedy Triangle, A Cloak for the Dreamer, The King's Commissioners,* and *Spaghetti and Meatballs for All*!

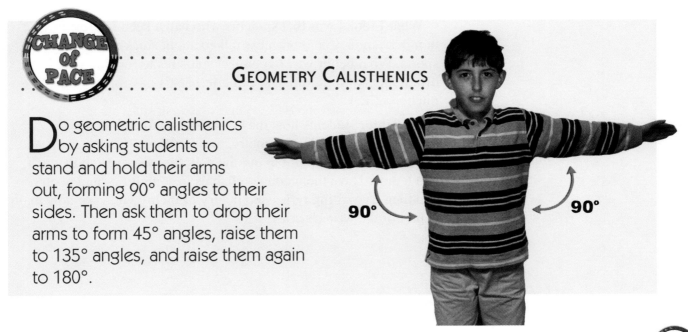

GEOMETRY CALISTHENICS

Do geometric calisthenics by asking students to stand and hold their arms out, forming 90° angles to their sides. Then ask them to drop their arms to form 45° angles, raise them to 135° angles, and raise them again to 180°.

90°   90°

Have You Heard Any Good Math Lately?

Listening for math is yet another highly effective way to help students connect math with the real world. It's also a great way to engage those students who are auditory learners.

Directions

When you ask students if they've heard any good math lately, you'll once again get that cross-eyed look. That's when you say, "Well, I heard the temperature today is going up to 75° and the low tonight will be about 32°. What's the difference?" Or ask, "Did anybody hear the score from last night's basketball game? The home team won, but by how many points?"

Suggest that students begin listening for sports scores, lottery prize amounts, odds for winning the lottery, temperatures, stock reports, time, amount of rainfall, and/or other statistics as they watch television, listen to the radio, or overhear conversations that include examples of math. Ask them to note and report on the math they hear. (Of course, you'll want to caution students to share only "public" numbers, and not personal or private conversations they might overhear.)

What I found was that students who had a keen interest in numbers, money, and business talked about stocks. Others talked sports. What better way to connect statistics to real life than by discussing batting averages, a quarterback's passing average, or a pitcher's ERA?

Model for students how they can make mathematical connections in music and poetry, too. Get them moving *and* learning as they clap or tap out the patterns, discovering the beat and tempo. Have them count syllables in song lyrics and poetry, and then try writing lyrics and poetry of their own to see how math is part of the writer's craft.

Have You Eaten Any Good Math Lately?

Start your students thinking creatively and mathematically by relating math to their morning breakfast. Then use Emeril's cooking philosophy of "kicking it up a notch." Help students internalize the concept of capacity by encouraging them to measure what they eat.

Directions

When you ask students if they've eaten any good math lately, they'll look at you cross-eyed once again. That's when you ask what they ate for breakfast that morning. At first, students will likely say that they had some cereal, eggs, some juice, perhaps doughnuts, or even nothing. As a teacher, at that point I liked to share what I ate for breakfast and mathematically model how I wanted the students to respond in the future.

MATERIALS

- Measuring cups
- Bowl
- Juice glass
- Cereal
- Juice
- Fruit
- Milk

I might start by cutting a grapefruit in half, explaining that I ate half a grapefruit. Next, I might pour some cereal into a bowl and say, "I ate about that much cereal. How can I figure out how much that is?"

I'd pour the cereal from the bowl into a measuring cup, and then I'd write on a little sheet of paper, "I had one cup of Cheerios." We would follow the same routine with the juice, pouring it into a glass and then into a measuring cup. Finally, I'd estimate how much milk I'd poured onto the Cheerios and measure that as well.

The goal is for students to internalize what four ounces of juice or one cup of cereal might look like. Once students have that sense of capacity, you can again ask them, "Have you eaten any good math lately?" This time a student might offer that she poured about one cup of cereal and four ounces of milk into a bowl, but ate only half that amount.

Using this strategy, you'll also learn something about your students' eating habits—especially when someone volunteers that his breakfast amounted to twelve ounces of root beer or even nothing at all.

Geometry Scavenger Hunt

Scavenger hunts like this geometry example help students reinforce their recognition and application of math concepts and vocabulary, while getting them up and moving around the classroom. This type of activity complements the earlier Warm-Up, "Have You Seen Any Good Math Lately?"

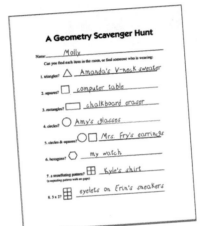

Preparation

Make a copy of the reproducible for each student or group in the class.

Directions

Give a copy of the scavenger hunt to each student. Explain that each child is to find an example of each item on the list and to write down what he finds. Encourage students to look for objects in the room and to examine what their classmates are wearing.

A child might find a square formed by a friend's pocket or a circle on the face of her own watch or the buttons on her jacket. She might identify the chalkboard eraser as a rectangle.

Notice example #8 on the scavenger hunt. This isn't really geometry, of course, but it *is* a review item that reinforces multiplication arrays. Students sometimes find examples in the room itself, but it's also enlightening when they find a 3 X 2 array in the eyelets of someone's sneakers or in the buttons on a double-breasted jacket or sweater.

Eventually, some students will start to identify representations— recognizing a triangle in a V-neck shirt or in the space between fingers spread in a "V for Victory" sign. The more students practice, the more proficient they'll become at finding all sorts of shapes and moving to higher-level thinking skills.

Now you can build on students' newfound knowledge of how

MATERIALS

- Reproducible #2 (or substitute another list that's appropriate for what the class has been studying)
- Ball of yarn or heavy string

Teacher Tip

Sometimes it can be hard for students to tell squares and rectangles apart. If that happens, invite them to measure each side, determine whether they've found a square or a rectangle, and explain the answer.

geometry applies to the real world. Begin this next step by creating a giant Venn diagram on the classroom floor, using yarn to make two overlapping circles.

Ask students to examine what they themselves—not their classmates—are wearing. Then say, "Everyone who's wearing circles, come stand in this circle. Everyone who's wearing squares, come stand in this other circle. And everyone who's wearing both, come stand in this section where the two circles overlap."

As they do this, students start to see similarities and differences, and they begin to master the concepts of geometric shapes. Searching for shapes in increasingly creative ways helps students leave the text and make sense of the math.

Variation

Use the same reproducible but tell students that, for each item, their task is to find as many examples as they can within a specified amount of time.

OTHER SCAVENGER HUNTS

The possibilities for mathematical scavenger hunts are endless, but include:

• **Math Facts Scavenger Hunt:** Each student walks around the room, asking everyone he meets, "Can you give me an equation for ___?" If his number is 8, answers might include 4 X 2, 10 – 2, or the square root of 64.

• **Measurement Scavenger Hunt:** Provide a list of measurements (e.g., 1 inch, 2 inches, 2½ inches), and challenge students to find objects in the room that match those measurements exactly. Make sure they actually measure the objects!

• **Phone Number Scavenger Hunt:** Give students pages from an old phone book and have them highlight all the phone numbers they can find with digits that add up to exactly twenty (e.g., 923-3120).

Overhead Operations

This interactive activity gives students practice with math facts, concepts, and vocabulary on a daily basis.

Preparation

Choose a number that will help you to reinforce concepts and skills the class has been studying. Count out that specific number of bingo chips.

MATERIALS

• Bingo chips

Directions

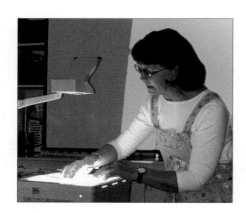

Explain to students that you're going to turn on the overhead for one second. Their objective is to estimate how many bingo chips they think they saw in that brief look, and write down the number. Tell them they'll have a chance to change their responses as you give mathematical clues for the exact number.

For example, you might place fifteen bingo chips on the overhead. Just put them in a pile, not in any particular order. Turn on the overhead for one second, then turn it off and ask each student to write down the number of chips he thinks he saw. Now give one clue at a time. Your comments might go something like this:

"We've been talking about properties of even and odd numbers. This is an odd number, so if you want, you can change your answer right now.

"We've been talking about prime and composite numbers. This is not a prime number, so if you want, you can change your answer right now.

"Remember how many items are in a dozen? This is greater than one dozen, so if you want, you can change your answer.

"We've been talking about square numbers, right? Well, I know that 4 squared is even, but this number is less than that. So if you want, you can change your answer.

"We've also been doubling numbers. What's the double of 7? It's greater than that.

"So what number do we have? That's right: 15."

This highly motivational activity reinforces math vocabulary and skills. Students are actively engaged, and everything is risk free, because you keep inviting students to change their answers. It's also infinitely adaptable. You know what the class or group has been studying and what you need to reinforce. All you have to do is choose a number that will let you reinforce target skills, vocabulary, and concepts, then give the appropriate clues.

Step 1

You can extend this activity, too. Let's say you've been working with fractions. Put those fifteen chips on the overhead, but this time arrange them in five rows of three. Say,

> "We have 15. What's 1/3 of 15? That's right: 5. Let's take away 5.
>
> "What do we have left? We have 2/3 of 15, which is 10.
>
> "So let's take away another third, which leaves 5."

Step 2

This hands-on, visual method helps many students to internalize the concept of fractions. Students love to come up to the overhead and remove the chips themselves, too.

Step 3

FUN ALTERNATIVES TO BINGO CHIPS

I love using bingo chips; they're colorful, they're easy to see on the overhead, and they work well as student manipulatives. I also like the fact that, because they have metal rims, they're easy to pick up with a magnetic wand.

But variety is good, too. It's fun to use other small manipulatives that you find in the confetti section of a party store. I jokingly say that this is curriculum integration at its best. On the first day of snow or winter, you might want to use snowflakes instead of bingo chips. If you're studying insects in science, you can use bugs. For holidays, you might use hearts, pumpkins, or shamrocks. They're not magnetic, so the confetti pieces are harder to pick up, but the students do enjoy using them.

Renaming the Date

The more students practice renaming the date, the more creative and proficient they become with numbers; this activity is a wonderful way to build number sense.

Preparation

Write the current day of the month on the board or on a transparency for the overhead projector.

Directions

Invite the class to create as many different equations for the day's number as they can in one or two minutes. Students can also rename the number by spelling the word or providing a foreign-language version of it. Give students one to two minutes to work on this individually, then ask for responses. As they volunteer their answers, add the responses to the transparency or the board.

September 10

5 X 2
100 – 90
1/3 of 30
"decade"
diez

The first time you ask students to rename the date, model the process for them. Suppose you're trying this Warm-Up for the first time on September 10. You'd explain that the goal is to rename the number 10.

Some possibilities for renaming would be: 5 X 2; 100 – 90; ten; 1/3 of 30. Someone might say that the word "decade" means ten years. Someone else might give the Spanish *diez*. You can always add anything they miss.

Once students are familiar with the activity, ask them to prepare their answers the night before this Warm-Up. This makes great tear-free homework! The next day in class, draw students' names at random (you might use craft sticks with one name on each). Ask each child whose name you draw to add to the list by "renaming the date."

Typically, what happens is that your higher-level students give you higher-level math. Others give you simpler answers. What's important is that everyone can participate, everyone can learn, and no one is wrong. It's a great way to reinforce equations.

Daily Number Review

We can't always rely on the textbook or math series to continually reinforce the language we want students to use. The Daily Number Review reinforces math vocabulary and number properties, and allows you to assess an individual student's mastery of these concepts.

Preparation

Keep an ongoing list of all the number properties the class studies. If appropriate, adjust the reproducible to match your class's needs (or the needs of a particular group) before copying

• Reproducible #3

it onto a transparency. Next, think of a number that will allow you to review as many items as possible from your list. It can be any number. It might be the current day of the month, or it could be a number telling how many days you've been in school so far this year. In the center box, write the number you've chosen for this day's practice.

Directions

If today is the thirtieth day of the school year, 30 might be the number for today's daily review. Suppose that, over time, the class has been studying these properties and terms: even and odd, prime and composite, palindromes, integers, counting/natural numbers, whole numbers, and square numbers. On the transparency you've made from the reproducible, you would begin by writing "30" in the box. From that point, the dialogue would continue along these lines.

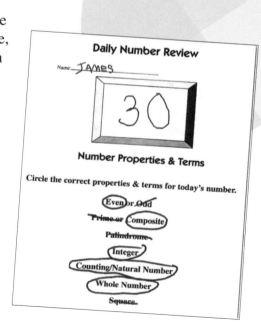

Teacher:	"Is this number even or odd?"
Students:	"It's even."
Teacher:	"Okay, so I'll circle 'even' and cross out 'odd.' Now, is 30 prime or composite?"
Students:	"It's composite."
Teacher:	"Why?"

Students:	"Because it has more than 2 factors."
Teacher:	"That's right. So I'll circle 'composite' and cross out 'prime.' Now, is it a palindrome? Does it read the same way backward and forward?"
Students:	"No."
Teacher:	"That's right. So I'll cross out 'palindrome.'"

You'd continue in this manner through the list—or whatever parts of it represent terms you expect your students to know at this point in the year. Be sure to ask for students' reasoning, not just their answers!

This is one more pathway that gives students a way to "play" with number properties and build number sense. Practice makes permanent, and the more purposeful practice we provide for students, the more likely it is they will understand, retain, and use the math they're learning.

Variations

Variation 1: Instead of working on the overhead, make a copy of the reproducible for each student. Write on the board the number you want students to work with, and have each child complete the work sheet on her own.

Variation 2: You can also differentiate by creating slightly different lists for students at different skill levels.

TELEVISION MATH

An effective homework assignment is to ask students to note the math they see on TV in an evening. I've seen them note numbers, shapes, and patterns, as well as starting, ending, and elapsed times. Some students will time the commercials and an actual program and show the ratio of advertisement time to program time. And they love being able to tell their parents that their homework assignment is to do math as they watch TV!

Term	Definition	Examples
Composite number	Has more than 2 factors; can be divided by numbers other than itself and 1	[4, 6, 8, 9, 10, 12, 14, 16, 18 . . .]
Counting numbers (also called natural numbers)	The set of numbers beginning with 1	[1, 2, 3, 4, 5, 6, 7...]
Even numbers	Can be divided exactly by 2. End in 0, 2, 4, 6, or 8.	0, 12, 24, 36, 464
Integers	Natural numbers and their negative counterparts, and 0	[-3, -2, -1, 0, 1, 2, 3, 4...]
Odd numbers	Cannot be divided exactly by 2; end in 1, 3, 5, 7, or 9.	1, 13, 25, 37
Palindrome	Reads the same forward and backward	212, 4004, 33, 1551
Prime number	Has only 2 factors: itself and 1	[2, 3, 5, 7, 11, 13, 17 . . .]
Rectangular number	Can be shown as a pattern of evenly spaced dots in the shape of a rectangle	2, 6, 8, 10, 12
Square number	Can be shown as a pattern of evenly spaced dots in the shape of a square	4, 9, 16, 25, 36
Whole numbers	Zero and the counting numbers (no fractions)	[0, 1, 2, 3, 4, 5, 6 . . .]

Note: 1 is considered unique.
It can be divided only by itself.

Number Stars

Use numbered stars to engage all students in an interactive review. This activity also gives you a chance to complete a quick skills assessment.

Preparation

You can make these yourself, or have students make them. Start on one side of a star, and write the number 1 on the first point, 2 on the second point, 3 on the third, and so on. Flip the star over and repeat with the numbers 6 through 10. Then go to another star and repeat the process for the numbers 11 through 20. Give a pair of number stars to each student.

MATERIALS

- Large stars from a party store (or cut star patterns from old manila folders)

Directions

I learned this strategy from another teacher, and found that it works really well. Since every student has her own stars, everyone participates. During a Warm-Up or Cool-Down, for example, you might say, "Show me the double of 10." No longer does one student raise her hand and say, "Oh, that's 20." Instead, everyone responds by showing her response under her chin.

Continue with other examples that are appropriate for your students; this can be anything from "show me 7 + 4" to "show me half of 32." Students get a chance to practice, and you get a chance to assess each child's level of learning.

Guess My Number

This strategy is a great way to reinforce math facts and get students up, moving about, and talking with each other.

Teacher Tip

If you want to be able to re-use the plastic plates, be sure to have the students write on them with dry-erase markers—not crayons. Crayon won't wipe off the plates. Or use paper plates and throw them away afterward or save them for art projects.

Preparation

Using a dry-erase marker, write a different number on each plate. Poke a hole in the top of the plate and attach a 14" length of yarn through the hole.

MATERIALS

- Dry-erase markers
- One snack- or dinner-size plastic plate for each student
- Yarn
- Music

Directions

Hang each plate around a student's neck so that the plate falls on his back and he can't see what's written on it.

Have the students walk around the room while you play music. Tell them that when you stop the music, they're to stop moving.

When the music stops, explain that each student should pair up with the person closest to him. Next, tell students that each of them is to write an equation on his partner's plate that equals the number on that plate.

Remind students that they're not to say anything—no clues or answers may be given! Then start the music again. Repeat this process as many times as you choose.

What might this look like? If the number on Amy's plate is 30, someone might write on it "5 X 6" or "29 + 1" or "300 ÷ 10." Amy's collecting as many ways of making 30 as she can. But she still doesn't know what's written on her plate.

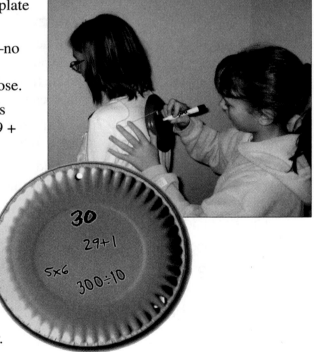

Stop the music one last time and ask students standing near each other to become partners and to read aloud the equations on each other's plates. Maybe Sara is standing next to Amy. She reads all the equations on Amy's plate, and Amy has to guess her own number. Then they repeat the process, with Amy telling Sara the equations on her (Sara's) plate and letting Sara guess her own number.

Thinkersize #1: Picture Puzzles

I want students to understand that when they play word games, they're using mathematical thinking. Many students say they don't like math, but they love reading, so I tell them that playing word games is a great example of using mathematical thinking.

Thinkersizes are great for developing analytical thinking and for helping students to recognize the role of patterns and relationships when looking for solutions. These kinds of puzzles can be good Warm-Ups.

Preparation

Copy the reproducible onto an overhead transparency.

MATERIALS

• Reproducible #4

Directions

Place the transparency on the projector, and cover all but the first puzzle, so everyone can focus on the same puzzle at the same time. Ask students to guess the answer, then model how they can figure out that it's "misunderstand."

Together, work out each of the remaining puzzles. You may need to help students come up with the answers.

Add more puzzles if you like; just be careful to use words or phrases that the children will recognize.

Follow up on this activity by inviting students to bring in their own examples to post on the bulletin board and share with the class. I've found that once they understand the concept, they become really good at both solving and creating word puzzles.

PICTURE PUZZLES ANSWER KEY

1. Misunderstand
2. Three degrees below zero
3. Book overdue
4. Walking on eggshells
5. Skating on ice
6. Walking on air
7. *The Cat in the Hat*
8. Right on time
9. Side by side
10. Man overboard
11. Something in my pocket
12. Once in a while

Thinkersize #2: Flexible Fred

This mental math activity helps students to improve their analytical thinking by identifying patterns of similarities and differences.

Directions

This is a concept-attainment form of learning, in which students attempt to figure out what attribute the examples have in common that the non-examples don't have.

Begin by saying to the class, "Flexible Fred likes Kermit the Frog, but not Miss Piggy. He likes grass, but not dirt. He likes summer leaves, but not flowers. What else might Flexible Fred like?"

You'll soon identify the analytical thinkers in your room. In this game, the objective is not to have students give you the rule right away. Instead, ask a more difficult question: ask them to give you new examples of things that Fred likes and doesn't like. This takes student thinking to a higher level.

Next, you go to the board or overhead and draw a T-chart with two columns—one for things Fred likes, and the other for things he doesn't like. The T-chart strategy gives students a visual, organized way to compare Fred's likes and dislikes to discover a pattern.

Now you want students to think of examples that will fit the rule. Someone might say, "He likes broccoli but not carrots." Eventually, you'll want someone to simply state the rule: Fred likes things that are green.

"Things that are green" is a good Flexible Fred to start with because it's not too difficult. Then, if you want to push things up a notch, you can say, "Flexible Fred likes desserts but he doesn't like pie. He likes candy but not sweets. He likes sneakers but not shoes. What else does he like and not like?"

If the students are stuck, create another T-chart, but add some hand prompts. Say, "He likes candy, desserts, and sneakers," tapping your chin twice as you say each word. "He doesn't like sweets, pie, or shoes,"

FLEXIBLE FRED	
Fred Likes	Fred Doesn't Like
Kermit	Miss Piggy
grass	dirt
summer leaves	flowers

tapping your chin *once* as you say each word. "What else might he like? Would he like trousers but not pants, sweaters but not shirts, a sofa but not a couch?" Again, tap your chin twice for trousers, sweaters, and sofa; once for pants, shirts, and couch. Somebody will figure out that Flexible Fred likes two-syllable words but not one-syllable words.

Do you see what we're doing here? We're practicing great mathematical thinking. Students are studying attributes, analyzing them to discover similarities and differences, and drawing logical conclusions.

Next, you might give this example: "Flexible Fred likes teenagers but not adolescents. He likes soccer but not sports. He likes glasses but not cups. What else does Flexible Fred like?"

If you've done the syllables comparison before this one, that's probably what students will look for now. Will it work? No. So students must come up with a different hypothesis and test it out. Draw the T-chart again, and someone will figure out that this time, Fred likes words with double letters.

You can find more "Flexible Fred" examples in the box on page 39, and once they get the hang of it, students will have fun coming up with their own. Ask them to show their examples to you before sharing them with the class, so you can check their hypotheses!

When I use Flexible Fred as a Cool-Down at the end of the day, I often take it one step further and use it as a line-up game. I'll say, "Okay, Jason, you're correct. For double letters, you gave me 'Mississippi but not Ohio.' So you can line up. And you can take Jim and Sarah and Matt with you."

I do that because it speeds things up, and because sometimes we have examples that not everyone will get. I know that, but I need to up the ante, and including difficult examples exposes everyone to some higher-level thinking. The important thing is that everybody improves.

1. Flexible Fred likes beach balls, but he doesn't like the beach.
   Flexible Fred likes sour balls, but he doesn't like licorice.
   Flexible Fred likes globes, but he doesn't like maps.
   What else might Flexible Fred like and not like?

   (things that are spheres)

2. Flexible Fred likes storms, but he doesn't like a hurricane.
   Flexible Fred likes doughnuts, but he doesn't like a muffin.
   Flexible Fred likes shoes, but he doesn't like a sneaker.
   What else might Flexible Fred like and not like?

   (words that are plural)

3. Flexible Fred likes boats, but he doesn't like the water.
   Flexible Fred likes cars, but he doesn't like roads.
   Flexible Fred likes buses, but he doesn't like streets.
   What else does Flexible Fred like and not like?

   (things that transport people)

4. Flexible Fred likes apples, but he doesn't like bananas.
   Flexible Fred likes artichokes, but he doesn't like squash.
   Flexible Fred likes anchovies, but he doesn't like fish.
   What else does Flexible Fred like and not like?

   (things that begin with the letter A)

5. Flexible Fred likes Sacramento, but he doesn't like San Francisco.
   Flexible Fred likes Harrisburg, but he doesn't like Pittsburgh.
   Flexible Fred likes Austin, but he doesn't like Houston.
   What else does Flexible Fred like and not like?

   (places that are state capitals)

Thinkersize #3: Hink Pinks

*H*ink Pink is actually a term that comes from old spellers. This terrific word-play activity reinforces mathematical thinking and reasoning by having students come up with rhyming synonyms to questions. Hink Pinks make a great Cool-Down exercise when you have just a few minutes at the end of the day.

Directions

Introduce Hink Pinks to your students by giving an example: "What's another name for an insect carpet? A bug rug." The answer to a Hink Pink is always two words, each with one syllable, and the two words need to rhyme. So an angry father is a "mad dad" and a gorilla coat is an "ape cape."

Once students understand the concept, copy onto the board any of the Hink Pinks from the box on page 41 and challenge students to solve them. It's important to have each student mentally flipping through the "Rolodex" in his head to come up with appropriate synonyms.

Once they catch on, ask students to invent their own Hink Pinks and bring them in. One of my students came up with "What's another name for 7 minutes after 5?" That's not an easy one! To figure it out, picture what we're talking about: a unit of measure related to time. Look at those two numbers, 7 and 5. Have you figured it out yet? It's "prime time." And that's mathematical reasoning.

If students become really good at it, they may also want to try coming up with a Hinky Pinky or even a Hinkity Pinkity. A Hinky Pinky has a two-word answer in which the words rhyme and each word has two syllables. A Hinkity Pinkity also has two words that rhyme, but now—you guessed it!—each word has three syllables.

What do you call a chubby kitten?

(fat cat)

For a Hinky Pinky, one of my students asked, "What's another name for a small numeral?" The answer: a midget digit.

Hinkity Pinkities are really tough! One of the best ones my students came up with was this: "What's another name for a happier dog? A merrier terrier."

HINK PINKS

CLUE	ANSWER
What do you call an unending melody?	long song
What do you call a damp animal?	wet pet
What do you call an insect embrace?	bug hug
What do you call a salmon plate?	fish dish
What do you call a chubby kitten?	fat cat
What do you call a termite house?	pest nest
What do you call an unhurried deer?	slow doe
What do you call a novel thief?	book crook
What do you call a rodent's headpiece?	rat hat
What do you call a tiny numeral?	wee three
What do you call wet postage?	damp stamp
What do you call a puppy kiss?	pooch smooch
What do you call a damaged penny?	bent cent
What do you call a filthy chicken?	foul fowl
What do you call a tired vegetable?	worn corn

Chapter 2

FOCUSING & RESPONSE TOOLS

Many students in our schools today have attention problems. The problem isn't that they can't pay attention; it's that they pay attention to everything. Some of these students are overwhelmed by the sheer number of problems on a page; others become confused when using a step-by-step process in multi-digit operations. We can help them focus by providing some inexpensive but effective highlighting tools.

Other tools included in this chapter help us as teachers to use time more effectively, and to conserve paper as well. In our classrooms, we continually assess student responses to check for understanding. But how do we check individual answers—or identify a student's thought processes—within a large-group situation? I've included some tools here that will do just that. They'll also help you to actively engage every learner in the class. I hope these ideas will be practical additions to your teaching toolbox.

One thing I want to emphasize is that all of these tools are designed to be used on an ongoing basis. They're not ideas to be tried once and then forgotten; they can help you and your students all year long.

Page Protectors

Page protectors are an all-time favorite of mine. They help you save paper, and they make set-up quick and easy. In addition, most page protectors are three-hole punched, so it's easy for students to store them in their binders.

Directions

Whenever you use a work sheet (such as a scavenger hunt, a daily number review, or a "mad minute") that has more than one correct answer and/or can be used again, all you need to do is put the sheet in a page protector. Students can use crayons, vis-à-vis markers, or dry-erase markers to complete the activity right on the page protector and correct it. Then they just wipe the page protector clean with a rag, and it's ready to go again.

CHANGE of PACE

A MNEMONIC TOOL FOR LONG DIVISION

Do you know the McDonald's mnemonic that students can use to remember the steps in long division? Tell the students to ask, "Does McDonald's sell cheeseburgers rare?" Then explain that *D* is for divide, *M* is for multiply, *S* is for subtract, and *C* is for check.

If everything checks out, bring down (from the *B* in burgers). They bring down the next number in the dividend and start the sequence again. Divide, multiply, subtract, check, bring down. Repeat this until there's nothing to bring down. Now say, "Do they sell those burgers rare? *R* stands for remainder."

Focus Frames

The focus frame is an excellent tool for keeping everyone on task and engaged in learning. Each student has a focus frame at his seat, and whenever it's appropriate, you instruct everyone to frame a vocabulary word or an answer to a question individually. Everyone participates. Then you walk around and observe what students have done.

Directions

Making focus frames is a great task for kids to complete on their own. Just give them old manila folders, scissors, and a pattern that you've created by cutting up one of the folders yourself. (That part's easy. To create the pattern, you just cut out two L-shaped pieces, then cut slits in one of them so the other piece can slide into it.)

MATERIALS

• Manila folders

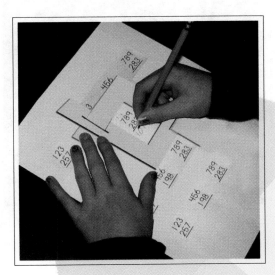

Students can use their focus frames in many ways. Most important, they frame, focus on, and solve arithmetic problems one at a time. So when a child pulls out his focus frame and puts it over a work sheet or his text, he knows he's going to work on this particular problem.

A focus frame helps in another way, too. Many students, especially those with motor coordination weaknesses or visual/spatial organization problems, tend to get numbers out of alignment. With the focus frame, students learn to frame and focus on units, then tens, then hundreds. They uncover what they need, one step at a time, to solve the problem. It's great for addition, subtraction, and multiplication, and it really helps in long division.

Envelope Windows

It doesn't get much easier than this. Ordinary window envelopes make great focus frames.

Directions

Take the front panel (the part that includes the window) from each of a bunch of window envelopes. Remove the acetate from each one. Give these to students and ask the students to frame and focus on specific words and numbers, or on one problem at a time.

MATERIALS

• Window envelopes

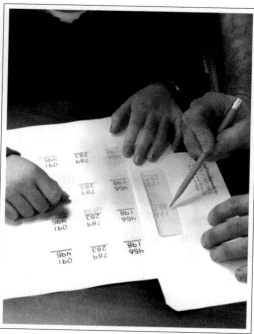

Highlighting Tape

Highlighting tape is reusable transparent colored tape. Students can use it to interact with text in much the same way adults use highlighting pens to mark the text they're reading.

Directions

When students learn "fix-up" reading strategies, such as identifying unknown words, highlighting tape provides an immediate way to mark the text so a child can come back and check the meaning of a word. Well, highlighting tape can also be extremely helpful in math.

The colored tape offers an alternate way for students to show their learning. Instead of asking a student to write a list, you might ask her to highlight all the fractions on a page that are not simplified to lowest terms. This helps you to determine if the student recognizes fractions in their simplest form before she moves on to the next level.

When students are working with word problems, highlighting tape can be an especially helpful tool. Have students highlight the question in one color and the numbers they need to solve the problem in another color. This strategy gives them both a process and a visual aid for approaching word problems.

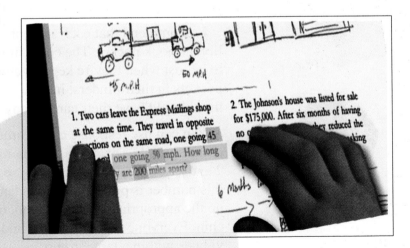

Response Cards

Response cards are a way to say, "Don't just raise your hand and give me the answer. Show me your thinking."

Directions

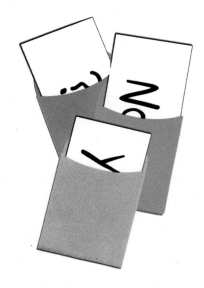

One of the problems we face in teaching is that we have students who are looking right at us but are not paying attention. Their minds are elsewhere. Typically, we ask questions and students raise their hands to answer. But when one student responds, the others immediately stop thinking. They're not engaged in learning.

MATERIALS

• 3 index cards (3" x 5") per student

How can we keep all students engaged and participating? By using response cards! No hands are raised; instead every student must respond, allowing teachers to monitor individual understanding within a large group.

To create response cards, give each student three 3" by 5" index cards. On one card he writes "yes," on the second he writes "no," and on the third he writes a question mark.

I recommend that each student store his response cards in a library book pocket. The cards fit perfectly, and they're less likely to be lost when they're kept together. Have students keep these pockets in their binders, in their desks, or on tables; explain that they should have their cards ready to use at all times.

You can use response cards throughout the day in all areas of the curriculum. For example, during math, you might show the number 2 and say, "Hold up your response card to show whether you think this number is prime." Each student chooses his response and holds up the appropriate card—yes, no, or question mark—under his chin. Everyone is accountable, and the teacher can assess individuals within a large group.

At this point, if I'm the teacher, I look out over the group and see students with yes, no, and question-mark responses. The next step is to ask questions: "Will someone with a 'no,' please explain?" Stephen might say, "Oh, I said no. Two can't be prime because it's an even number."

Now I know why Stephen said no. Then I see that Aaron has a "yes." I ask him to explain his response and Aaron says, "Even though two is even, it only has two factors." Other students can change their responses, based on this new information. Stephen might still think two can't be prime and keep his "no" card up. On the other hand, if he understands Aaron's point, he changes to the "yes" card.

But it's the question mark that can really help. I'll say, "Of those who aren't sure, who would like to share why?" I might have a child say, "Tell me again what a prime number is." That answer indicates the student isn't sure. I'll also have students who will say nothing. That's when I go up to each of them privately and say, "I noticed you held up a question mark. How can I help you?"

If the student asks me to repeat the question, I know I've identified someone who may be looking directly at me, but is not paying attention. Today we must add "tuning in" to the skills we teach. Frequent use of response cards helps learners tune in.

. . . .
RULES WHEN USING RESPONSE CARDS
. . . .

- Students may not ask each other how to respond.
- Students may not say anything aloud, such as, "Oh, of course it's prime!"
- Students may not look at others' responses.

Dry-Erase Tools

Dry-erase markers and boards are wonderful tools, but if they don't fit within the budget, substitute large plastic picnic plates.

Directions

You can use the plates in much the same way you use response cards, but the plates allow for a wider range of specific responses. You might say, "Show me a way to represent the number 7." One student might write "6 + 1" and another might write that it's the square root of 49. You can quickly assess the different levels of their thinking.

LEARNING MATH THROUGH PATTERNS

There are three tools I recommend students use throughout the year: a Hundred Chart, an Addition Table, and a Multiplication Table. These are highly effective tools because they help students identify patterns and relationships. Patterns are the key, I believe, to building student understanding and mastery of important math concepts.

You'll find reproducibles for the Hundred Chart (#5), Addition Table (#6), and Multiplication Table (#7) in the back of this book. I suggest that you make a copy of the Hundred Chart for each child in your class and place it in a page protector for easy, see-through storage.

You can do exactly the same thing with the Addition and Multiplication Tables. It's handy to photocopy one of these onto the front and the other onto the back of the same sheet of paper, then place that sheet in another page protector. This allows a child to work with the Addition Table, then just flip the page protector over to use the Multiplication Table.

I've included triangular flash cards in this chapter because they also reinforce student understanding of patterns and relationships. And they're great for practicing at home and building math fact automaticity.

Sum Patterns

This activity reinforces students' understanding of number families and the commutative property, which helps build math fact automaticity and computational fluency.

Preparation

Setting the enlargement feature on the copier to 105% to allow more space for the bingo chips, make one copy of the reproducible for each student, plus one overhead transparency for yourself.

MATERIALS

- Reproducible #5 (Hundred Chart)
- 2 sets of bingo chips for the overhead projector, each in a different color

Directions

Let's consider the importance of fluency or automaticity in learning. Do you drive to school by pretty much the same route every day? Are you "on automatic"? Have you ever arrived at school and had absolutely no memory of the trip? Now *that* can be a little scary!

Still, automaticity can be helpful, because the more you operate "on automatic," the more your mind is freed up to do other things, such as think about what you need to do at home or work. It's the same with school. For reading, a child needs to decode at an automatic level or he has difficulty comprehending. He's so busy trying to figure out how to say the words that it's hard to think about what they mean.

In math, if you're trying to teach multi-digit addition or long division or algebra, and the child is still at the manual stage of adding 2 plus 3, there's a problem. That child may be going through all the motions—counting on his fingers, etc.—but he doesn't have automaticity.

We want students to build automaticity with math facts. This Hundred Chart strategy works well for students who are beginning to learn addition facts or for those who have not yet discovered the patterns inherent in these math facts. It helps them to visualize and to build up that "mental Rolodex" of long-term memory.

Place the transparency of the Hundred Chart on the overhead projector and pull out your two colors of bingo chips. Proceed as follows:

• On the top row of the chart, cover the 2 with a bingo chip. Let's say you use a blue chip.

• Move down diagonally one square to the left. You are now on the 11. Cover the 11 with a chip in a different color. Let's say that one is red.

• Return to the top row and cover the 3 with a blue chip.

• Move downward and diagonally to the left. Cover the 12 and the 21 with red chips.

• Move to the top row again. Cover the 4 with a blue chip.

• Move down diagonally to the left, covering the 13, 22, and 31 with red chips.

Continue with this pattern, covering the number in the top row with one color, then the numbers in the left diagonal with a second color. The dialogue might go something like this:

Multiplication Table

Teacher:	Do you see a pattern here? Does anybody notice anything about the numbers that have red chips and the numbers that have blue chips? If you look at the 11, what do you notice?
Students:	The two digits of the 11 add up to 2.
Teacher:	That's right. Now let's study the 3. What do you notice about the numbers 21 and 12 in the rows below the 3?
Students:	The 2 and the 1 add up to 3, and the 1 and the 2 add up to 3.
Teacher:	That's right. Now let's look at the 4. What do you notice about the 13, the 22, and the 31?
Students:	The digits always add up to 4.
Teacher:	That's right. And does it matter which order you add them? If you add 1 + 3, do you get the same answer as if you add 3 + 1?
Students:	It doesn't matter.
Teacher:	That's right. That's called the commutative property. It means that no matter what numbers you're

adding, the order doesn't matter. You can add 2 + 1 or 1 + 2, and you'll get the same answer. What else can you add that will give you the same answer?

Students: 1 + 4 and 4 + 1 [or any other number combinations that show up on the diagonal of the Hundred Chart].

From there, you continue in the same vein. Seeing these patterns empowers students; the patterns give them a visual, concrete way to internalize addition facts.

SOURCES FOR BINGO CHIPS

I can't imagine teaching without bingo chips. You've probably already figured out that I use them all the time to illustrate math concepts and patterns, and I love working with them. So do students; they can use them at their seats or in centers for fun learning *and* clean and easy pick-up.

If you have trouble finding bingo chips, try looking in teacher stores.

Lakeshore Learning Materials carries them in their stores and catalogues, too. Sometimes you can find the magnetic wands in fabric stores. It's worth seeking them out.

Triangular Flash Cards

Another way to reinforce automaticity with number families is to have students work with triangular flash cards. Trend and ETA/Cuisenaire sell these ready-made, or you can have students make their own sets.

Directions

Give students a stack of old manila folders and have them cut out equilateral triangles that measure roughly 4 or 5 inches on each side. (I recommend rounding the edges so the cards don't get dog eared.)

MATERIALS

• Manila folders

Next, in the corners of the cards, students write the numbers that are appropriate for the fact families they're studying. For a student studying sums to 8, the process goes something like this:

• Using a blue marker, the student writes an 8 at the top of each of 4 flash cards. Then she turns the cards over and does the same thing on the backs.

• She takes one card and, using a green marker, writes a 1 in the bottom left corner and a 7 in the bottom right corner.

• She turns the card over. Still using a green marker, she writes a 7 in the bottom left corner, then a 1 in the bottom right corner. This double-sided flash card demonstrates the commutative property, that 1 + 7 is the same as 7 + 1.

• She takes the next card and writes a green 2 and a green 6, in that order, in the bottom corners. Then she flips the card and writes a green 6 and a green 2, in *that* order, in the bottom corners.

• On the third card she writes a green 3 and a green 5 in the bottom corners, then flips the card and writes a 5 and a 3, both still in green.

• On the last of the "sums to 8" cards, she puts a green 4 in each of the 4 bottom corners (two on each side) of the card.

Once students create their own triangular flash cards, they can practice solving for the missing number in the fact family. All they have to do is cover up one of the three numbers and ask, "What's the missing number in the family?"

For example, if a student has a card with an 8, 1, and 7 on it, he covers the 7 with his thumb and asks, "What's the missing number in this family?" It's 7. The student self-checks by lifting up his thumb to find the 7.

You can develop mathematical vocabulary and understanding of concepts with these cards, too. Initially, ask students to identify missing numbers in fact families. Later, ask them to identify these missing numbers as missing addends. Still later, ask them to solve for X.

These flash cards can be used at home and at school throughout the year. They provide one more pathway for a child to learn math facts.

Students can also create triangular flash cards for multiplication and division facts. For that purpose, the triangular flash cards might look like the one at the left.

With the multiplication cards, instead of solving for the missing addend, students are now solving for the missing factor. The principle is exactly the same: you're teaching fact families. You're also reinforcing the concept that multiplication and division are related, just as addition and subtraction are related; they're inverse operations.

When students construct their own triangular flash cards, they reinforce their understanding and learning. They also have ownership of their cards.

Adding & Subtracting Two-Digit Numbers

This is an excellent hands-on approach for students who have learning difficulties associated with addition and subtraction. It's also a method that all students can use to check their own addition and subtraction computations. One cautionary note: It works only with single- and two-digit numbers.

Preparation

If you haven't already done so for previous activities, make one copy of the reproducible for each student, setting the enlargement feature of the copier to 105%.

MATERIALS

- Reproducible #5 (Hundred Chart)
- 2 different-colored bingo chips per student

Directions

First guide an individual or group of students through the process. Give the student a copy of the Hundred Chart and two bingo chips in different colors. Let's say the student wants to add 34 plus 22.

- On the Hundred Chart, have the child begin by placing one bingo chip on the 34.

- Now, he needs to add 22. To do this, have him take the other bingo chip and count by 10s down the column, moving to 44 and then 54.

- He has now added 20 to 34 and is on 54.

- Have him add the 2 units by counting horizontally to the right: 55, 56. The sum is 56.

Hundred Chart

1	2	3	4	5	6	7	8	9	10
11	12	13	14	15	16	17	18	19	20
21	22	23	24	25	26	27	28	29	30
31	32	33	34	35	36	37	38	39	40
41	42	43	44	45	46	47	48		
51	52	53	54	55	56	5			
61	62	63	64	65					
71	72	73	74						
81									

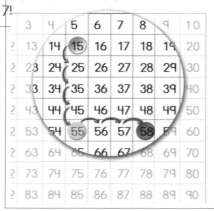

Let's try another example, an easy one. Let's say you ask a student, "What's 15 plus 43?"

• The student places the first chip on 15.

• He uses the second chip to add 40, counting down the column by 10s. He pauses on 25, 35, and 45, then stops on 55.

• From there, the child counts to the right on that same row and adds the 3 units, stopping on the total sum of 58.

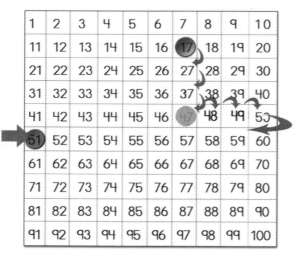

Once the student understands these 2 examples, you can move on to something a little more difficult: an example in which the student needs to regroup. Let's say the child needs to add 17 plus 34.

He starts with the 17 and counts down the column by tens: 10, 20, 30. He pauses on the 27 and 37, and stops on the 47.

But when he goes to add the 4 units by counting to the right, he runs out of squares in that row. That's when he needs to continue on to the next row. So for that equation, the student would work like this: He would count on from the 47 to 48, 49, 50, then sweep down to the beginning of the next row and stop on 51.

Subtracting with the Hundred Chart is similar. You simply reverse the process used for addition. Let's try the example of 68 - 32.

• With subtraction, you start with the minuend (greater number). So have the student put one chip on the minuend, 68.

• The subtrahend (lesser number) is 32, so have the child subtract 30 by counting by 10s *up* the column, pausing on 58 and 48, then stopping on 38.

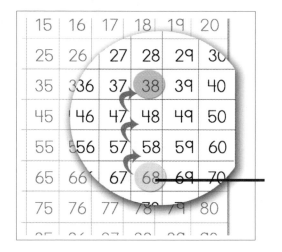

minuend

• To subtract the 2 units, have the student count 2 squares to the left of the 38, ending up on the answer: 36.

• This process empowers different students for different reasons. But it's especially helpful for struggling learners, and that's why I like to include it in my instructional strategies.

Multiple Patterns & Finding Common Multiples

U se this concrete, pattern-based strategy when you're teaching skip counting to younger students or multiples to older ones.

Preparation

If you haven't already done so for previous activities, make one copy of the reproducible for each student, and one overhead transparency. I suggest setting the enlarging feature of the copier to 105% for this, to allow more room for the bingo chips.

MATERIALS

- Reproducible #5 (Hundred Chart)
- 20 bingo chips in one color for each student
- 20 bingo chips in a second color for each student

Directions

I recommend you demonstrate this on the overhead transparency as students work along at their seats. Give each student a copy of the Hundred Chart and 20 bingo chips in each color (a total of 40 chips per student). Then model for students what happens when you skip count by twos, which of course is the same as finding the multiples of 2.

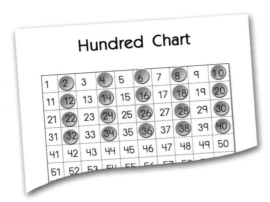

Hundred Chart

- Begin by placing a chip on the 2. Let's say it's a blue one.

- Count by twos, placing a blue chip on every other number: 2, 4, 6, 8, 10, 12, 14, 16, 18, 20, 22, 24, 26, 28, 30, and so on.

- Ask students what pattern is emerging on the Hundred Chart.

- They'll note that straight lines or columns are forming.

- When working with older students and talking about multiples, add another step to this activity. Have students look at the pattern, then make predictions. You might ask, "Is 45 a multiple of 2?" Students should respond, "No, because it doesn't fit the pattern." Then you ask, "Is 76 a multiple of 2?" Students should see that the answer is yes, because 76 fits the pattern.

Students can find patterns for every set of multiples on the Hundred Chart. Here's what happens with multiples of 3. Have the students place chips—let's say these are red ones—on 3, 6, 9, 12, 15, 18, and 21. Ask if they see a pattern emerging. Someone will recognize that the multiples of 3 form diagonals on the Hundred Chart.

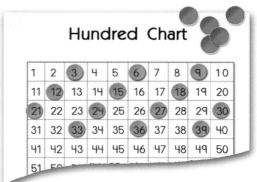

Ask students to continue the pattern, as you pose prediction questions similar to those you asked when working with multiples of 2: "Is 45 a multiple of 3?" "Is 26 a multiple of 3?" Without doing the computation, students can see whether or not a number is a multiple, based on the pattern.

Try bumping this up a notch by looking for common multiples. Let's say we've found the multiples of 2 and 3 and the students can visualize those parallel lines for the 2s and the diagonals for the 3s. Now they're trying to find common multiples.

• Have students start over again and mark the multiples of 2 with blue chips. They should place blue chips on 2, 4, 6, 8, 10, 12, 14, 16, 18, 20—all the way up to 40.

• Have students place red bingo chips on the multiples of 3. So they mark 3, 6, 9, 12, 15—this time, up to 39.

• Tell students that if they come to a number that already has a blue chip on it because it's a multiple of 2, they should place the red chip directly on top of the blue one; that number is a multiple of both 2 and 3.

• When students have marked all the multiples of 2 and 3, ask them what they notice about their common multiples. Students will see that the blue and red chips, when stacked on top of each other, turn purple. The common multiples are visibly purple.

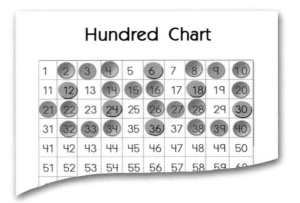

This hands-on visual strategy once again helps students find and use patterns. In contrast to simply reading about common multiples in a text, this approach invites students to discover the meaning of common multiples in a way that helps them to store the learning in long-term memory.

Finding Averages Through Patterns

This approach teaches the use of patterns as well as the problem-solving strategy of working backward. I've found it's a meaningful way to introduce averaging to students.

Preparation

If you haven't already done so for previous activities, make one copy of the reproducible for each student, and one overhead transparency for yourself.

Directions

Place the transparency of the Hundred Chart on the overhead. Explain to students that you're going to ask them to do some computation, as well as to look for a pattern. Then proceed as follows:

• Place a bingo chip in the same color on each of these numbers on the chart: 1, 3, 21, 23. Let's say you use blue chips.

• Ask students to find the sum of the 4 values. They should tell you it's 48.

MATERIALS

• Reproducible #5 (Hundred Chart)

• Bingo chips in 2 colors

• Calculator for each student

• Ask the students to divide 48 by 4 (the number of values you added). They should give you a response of 12.

• Cover the 12 with a bingo chip in a different color—let's say red. This forms a pattern.

• You want students to see the pattern that results from this averaging activity. Be sure they understand that, in this example, the 12 on the Hundred Chart represents the average value of the set of values: 1, 3, 21, and 23.

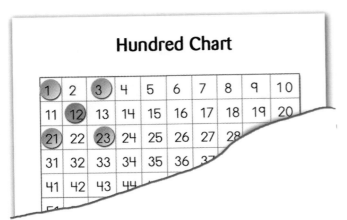

Hundred Chart

1	2	3	4	5	6	7	8	9	10
11	12	13	14	15	16	17	18	19	20
21	22	23	24	25	26	27	28		
31	32	33	34	35	36	37			
41	42	43	44						

• Now, of course, you want the students to predict an average. Place your blue bingo chips on the numbers 1, 5, 21, and 25. Ask students to predict the average of those 4 numbers, based on the pattern. They should come up with 13. Mark the 13 with a red chip.

• What do you tell the students to do next? You tell them to prove it.

• With the students, demonstrate computing the average. Add the values 1 + 5 + 21 + 25, then divide by 4 (the number of addends) to prove that the average is 13.

• Ask students to create averaging problems of their own by predicting (using their copies of the Hundred Chart) and then proving. Students can use calculators to prove their answers if they create averaging problems that are more advanced than their computation skills.

I like to tell students that they can find a lifetime of averaging problems on the Hundred Chart. Give them plenty of opportunities to practice by creating their own problems both in class and as homework. Some students—usually those who don't like text or workbook examples—actually look forward to creating their own averaging problems.

This strategy also provides another, more important benefit. Later in the year, when students encounter averaging problems in the book or on a test, they can reflect back to that concrete, visual experience of using patterns to find averages.

WANT TO KNOW A SECRET?

The secret to using the Hundred Chart to average 4 numbers is this: whenever you create a square or a rectangle from a section of the Hundred Chart that has an odd number of squares on each side, the middle value will be the average of the values in the corner squares. But it works *only* with odd numbers of squares.

If a student attempts to find a middle value when she creates a square or rectangle with an even number of squares on each side, she'll discover on her own that it doesn't work. She can certainly compute the average of the four corners, but she won't find a center value that shows the average on the Hundred Chart.

Focusing on the Facts

Focusing on the Facts is an activity that helps students focus their attention on number families. It involves the use of color strips to reinforce the relationship between addition and subtraction facts on the Addition Table and the relationship between multiplication and division facts on the Multiplication Table. Using the color strips with the tables also allows students to self-check their math facts when playing games or doing single-digit math computation.

Preparation

Make two-sided paper copies of the reproducibles for each student, with the Addition Table on one side and the Multiplication Table on the other. (I recommend reproducing each of these at 105%. The slightly enlarged copies will accommodate other activities.) Then make one overhead transparency of each reproducible. Give one paper copy to each student, and have each child insert his sheet in a page protector.

MATERIALS

- Reproducible #6 (Addition Table)
- Reproducible #7 (Multiplication Table)
- 1 page protector for each student
- Acetate report covers in 2 or more colors

Next, either you or your students need to cut strips from acetate report covers in two different colors, making sure that each strip is approximately the same width and length as the columns in the Addition Table. Each student can then store a set of these strips with his chart inside his page protector.

Directions

Many textbooks include addition and multiplication tables. These charts are math aids, meant to help students. But for many students, the charts aren't helpful because they're too busy. You can solve this problem with colored acetate strips.

Place your transparency of the Addition Table on the overhead, and invite students to follow your modeling with their own materials at their seats. Pull out one acetate strip in one color—let's say red—and one in another color—let's say green. Now, model how you can solve the problem 2 + 3.

• Place the red strip horizontally over the row that begins with 2.

• Place the green strip vertically over the column that begins with 3.

• Ask students to focus on the number where the 2 strips overlap. The strips highlight the answer, which is 5. The colored acetate strips allow students to focus on the facts, reinforcing their understanding of the number families.

Using the color strips also helps students to see the commutative property and to recognize that addition is the inverse operation of subtraction. The strips on the chart reinforce these concepts in much the same way the triangular flash cards (see page 55) do—by providing a concrete visual representation. By using the strips and the chart, students are able to see that:

| | Addition 1 |

0	1	2	3	4	5	6	7
1	2	3	4	5	6	7	8
2	3	4	5	6	7	8	9
3	4	5	6	7	8	9	10
4	5	6	7	8	9	10	11
5	6	7	8	9	10	11	12
6	7	8	9	10			

• 2 + 3 = 5   • 3 + 2 = 5

• 5 − 2 = 3   • 5 − 3 = 2

You can use the Multiplication Table in the same way. Focusing on the facts allows students first and foremost to see the relationship between multiplication and division. Division is the inverse of multiplication, and vice versa. Using the chart and color strips helps reinforce student recall and use of multiplication and division facts.

The chart and strips also allow students to check their answers independently during math games and activities. For example, if a student wants to check his answer to the problem 3 times 5, he can place a strip in one color on the column that begins with 3, and a strip in another color on the row that begins with 5. The colors overlap on the answer: 15. The student can also see that 15 divided by 5 is 3, or that 15 divided by 3 is 5. The strips highlight the relationship.

Multiplication Table

1	2	3	4	5	6	7
2	4	6	8	10	12	14
3	6	9	12	15	18	21
4	8	12	16	20	24	28
5	10	15	20	25	30	35
6	12	18	24	30	36	42

Skip Counting & Multiples

This is a visual, concrete way to introduce students to the concept of multiples by using the Multiplication Table and colored acetate strips. Go back to this tool over the course of the school year to introduce and reinforce appropriate multiples. Younger students can use the chart for skip counting.

Preparation

Make one copy of the reproducible for each student, and one overhead transparency. Then cut acetate strips (or have students cut them) from report covers, making each strip the same width as the rows in the Multiplication Table.

MATERIALS

• Reproducible #7 (Multiplication Table)

• Colored acetate report covers

Directions

Pull out your transparency of the Multiplication Table, along with a colored acetate strip. Ask students to watch as you demonstrate, then invite them to use their own charts and strips to follow along. Start by placing the color strip on the column that begins with 2. Point out that the strip highlights the multiples of 2. Discuss the properties shared by numbers that are multiples of 2.

Next, use a strip to highlight the multiples of 3. Continue helping students recognize that as they learn multiples of numbers 1 through 12, they can highlight them on the Multiplication Table. The color highlights the pattern of the multiples, which makes the concept easier for many students to understand.

Once they recognize multiples, you can show students how to use the chart to find common multiples (see the next activity).

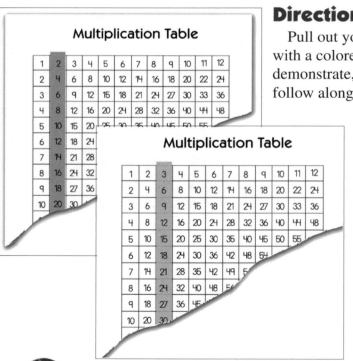

Another Way to Find Common Multiples

This method of using the Multiplication Table builds on the previous activity, in which students learned to identify multiples of numbers. Here they find common multiples (or denominators) in a concrete, visual way before working with a computational method.

This process helps students gain a solid understanding of common multiples, so that they can move on to the abstract concept of finding the least common multiple (also known as the least common denominator). They can then learn to create like fractions, and to add or subtract those fractions. For students who have difficulty abstractly finding the least common multiple, this tool is invaluable.

Preparation

If you haven't already done so for previous activities, make one copy of the reproducible for each student, and one overhead transparency. Cut acetate strips (or have students cut them) from report covers, making each strip the same width and length as the columns in the Multiplication Table.

MATERIALS

- Reproducible #7 (Multiplication Table)
- Acetate report covers in 2 colors

Directions

To demonstrate the process of finding common multiples, take your transparency and acetate strips and proceed step by step.

• Identify and highlight the multiples of two numbers. If the numbers are 4 and 6, place a strip in one color over the 4s column and a strip in another color over the 6s column.

• Go to the greater number and start moving down that column. In this case, ask the class, "Is 6 a multiple of 4? No. Is 12 a multiple of 4? Yes."

Multiplication Table

1	2	3	4	5	6	7	8	9	10	11	12
2	4	6	8	10	12	14	16	18	20	22	24
3	6	9	12	15	18	21	24	27	30	33	36
4	8	12	16	20	24	28	32	36	40	44	48
5	10	15	20	25	30	35	40	45	50	55	60
6	12	18	24	30	36	42	48	54	60	66	72
7	14	21	28	35	42	49	56	63	70	77	84
8	16	24	32	40	48	56	64	72	80	88	96
9	18	27	36	45	54	63	72	81	90	99	108
10	20	30	40	50	60	70	80	90	100	110	120
11	22	33	44	55	66	77	88	99	110	121	132
12	24	36	48	60	72	84	96	108	120	132	144

• The class has just found, by using patterns on the chart, that the least common multiple for 4 and 6 is 12. You can teach students that they can multiply two numbers to get a common multiple, but they need to know that this doesn't necessarily mean they're identifying the *least* common multiple. In the case of 4 and 6, students would multiply to get a product of 24. This certainly is a multiple of both numbers, but it's not the lowest one.

Once you've modeled the process, have students try it on their own. Then have them practice—frequently. You might even include this as a Warm-Up or Cool-Down activity. Choose two numbers and ask questions such as "What's the least common multiple for 3 and 5? Is it 5? 10? 15?"

With guided practice like this, students improve their comfort level and their ability to identify and work with common multiples. And that eases the transition to finding the least common denominator when working with unlike fractions.

Equivalent Fractions

This approach to finding equivalent fractions makes learning far more developmentally appropriate for students than traditional text-book explanations.

Preparation

If you haven't already done so for previous activities, make one copy of the reproducible for each student, and one overhead transparency. Cut acetate strips (or have students cut them) from report covers, making each strip the same width and length as the columns in the Multiplication Table.

MATERIALS

- Reproducible #7 (Multiplication Table)
- Acetate report covers in 2 colors

Directions

For many students, equivalent fractions are confusing and abstract, but when you take the Multiplication Table and systematically apply color to it, the concept of equivalent fractions comes alive. The process goes like this.

• Place the transparency of the Multiplication Table on the overhead.

• Place one of the colored acetate strips on the row that begins with 1. Place a strip in a different color on the row that begins with 2.

• Tell students that when they look at this chart, you don't want them to think of rows. Instead, they should think of fractions.

• Ask, "Does everyone see 1/2 on the chart? Do you see the other fractions in that same row: 2/4, 3/6, 4/8? Would someone like to make a hypothesis about these fractions? What do you think? Might they all mean the same thing? That's right. When we look at these two rows as fractions, they're all equivalent to 1/2."

Multiplication Table

1	2	3	4	5	6	7	8	9	10	11	12
2	4	6	8	10	12	14	16	18	20	22	24
3	6	9	12	15	18	21	24	27	30	33	36
4	8	12	16	20	24	28	32	36	40	44	48
5	10	15	20	25	30	35	40	45	50	55	60
6	12	18	24	30	36	42	48	54	60	66	72
7	14	21	28	35	42	49	56	63	70	77	84
8	16	24	32	40	48	56	64	72	80	88	96
9	18	27	36	45	54	63	72	81	90	99	108
10	20	30	40	50	60	70	80	90	100	110	120
11	22	33	44	55	66	77	88	99	110	121	132
12	24	36	48	60	72	84	96	108	120	132	144

• Have students place one color strip on the row beginning with 2 and the second color strip directly below it on the row beginning with 3. Be sure they now see these two rows as fractions, beginning with 2/3. Move from left to right, having students identify 4/6, 6/9, 8/12, and so on. Help them to see that in this pair of rows, the fractions are equivalent to 2/3.

• Move the strips again, this time placing one on the row beginning with 3, and the other on the row beginning with 4. (See Figure 1.) Move from left to right, helping students to discover the equivalent fractions for 3/4.

• Continue this pattern, showing equivalent fractions for each set of the adjacent rows: 4/5, 5/6, 6/7, 7/8, 8/9, 9/10, 10/11, and 11/12.

• Give students a chance to practice finding equivalent fractions on their own, using their acetate strips and individual copies of the Multiplication Table.

Multiplication Table

1	2	3	4	5	6	7	8	9	10	11	12
2	4	6	8	10	12	14	16	18	20	22	24
3	6	9	12	15	18	21	24	27	30	33	36
4	8	12	16	20	24	28	32	36	40	44	48
5	10	15	20	25	30	35	40	45	50	55	60
6	12	18	24	30	36	42	48	54	60	66	72
7	14	21	28	35	42	49	56	63	70	77	84
8	16	24	32	40	48	56	64	72	80	88	96
9	18	27	36	45	54	63	72	81	90	99	108
10	20	30	40	50	60	70	80	90	100	110	120
11	22	33	44	55	66	77	88	99	110	121	132
12	24	36	48	60	72	84	96	108	120	132	144

Figure 1

Students can also use this approach to find equivalency for fractions with numerators and denominators not in adjacent rows. For example, if a student wants to find equivalent fractions for 5/8, she can put one color strip on the row that begins with the numerator of 5. Then she can place the other color strip on the row that begins with the denominator of 8. (See Figure 2.) Even though the rows are not touching, the color shows the pattern with fractions that are equivalent to 5/8.

Multiplication Table

1	2	3	4	5	6	7	8	9	10	11	12
2	4	6	8	10	12	14	16	18	20	22	24
3	6	9	12	15	18	21	24	27	30	33	36
4	8	12	16	20	24	28	32	36	40	44	48
5	10	15	20	25	30	35	40	45	50	55	60
6	12	18	24	30	36	42	48	54	60	66	72
7	14	21	28	35	42	49	56	63	70	77	84
8	16	24	32	40	48	56	64	72	80	88	96
9	18	27	36	45	54	63	72	81	90	99	108
10	20	30	40	50	60	70	80	90	100	110	120
11	22	33	44	55	66	77	88	99	110	121	132
12	24	36	48	60	72	84	96	108	120	132	144

Figure 2

Adding & Subtracting Unlike Fractions

Once students understand how to use the Multiplication Table to find equivalent fractions (see previous activity), you can show them how to use it to add and subtract unlike fractions. This is an especially helpful approach for students who have difficulty with fractions.

Before beginning this strategy, it's important to be sure students already understand the concept of like and unlike fractions. They should also know how to add and subtract like fractions. This is an alternative way for students to add and subtract fractions with unlike denominators.

The table approach works especially well for students who are having a hard time computing abstract math. It allows them to "come to the party" along with everyone else. They just use a different pathway to get there.

Preparation

If you haven't already done so for previous activities, make one copy of the reproducible for each student, and one overhead transparency. Cut acetate strips (or have students cut them) from report covers, making each strip the same width and length as the columns in the Multiplication Table.

MATERIALS

- Reproducible #7 (Multiplication Table)
- Acetate report covers in 2 colors

Directions

Before introducing this new strategy, review the concept of like fractions (fractions with the same denominator). Ask students to

As always, it's important to model this strategy, going step by step with the students. Then follow the demonstration with guided practice that's appropriate and ongoing.

add and subtract some examples of computation problems with like fractions.

Next, review your lesson on unlike fractions (fractions with different denominators). Ask students to share what they learned about adding and subtracting unlike fractions. Be sure to have students verbalize and demonstrate the concept that before adding or subtracting unlike fractions, you must find a common denominator.

Now you're ready to demonstrate this strategy. To model for students how to find the least common denominator and then add and subtract unlike fractions, follow these steps:

• Put the Multiplication Table on the overhead projector, and pull out two acetate strips in different colors. Explain that you're going to share with the class how to use patterns on the chart to add and subtract unlike fractions.

• As an example, you might write on the overhead projector this problem: 1/7 + 2/5 =

• Ask students if we can add these fractions. They should say we can't because 1/7 and 2/5 are unlike fractions (they have different denominators).

• Confirm that the students are correct, and that we can add or subtract only like fractions. "So," you might ask, "how do we take 1/7 and 2/5 and convert them to like fractions? We find the least common multiple or denominator!"

• Invite the students to take out their Multiplication Tables to find the least common denominator for 5 and 7. From previous practice, students should place one color strip on the 5s column and the other on the 7s column. Then, starting with the greater number (7), they should cross-reference the multiples until they find a common one for 7 and 5. It's helpful to remind students to use a questioning process (e.g., Is 14 a multiple of 5? No. Is 21? No. Is 28? No. How about 35? Yes!) They should recognize that 35 is a multiple of both 7 and 5. It's the least common multiple, or the least common denominator.

• Explain that we now know that 35 is the least common denominator, but now we must convert 1/7 and 2/5 to equivalent fractions, each with a denominator of 35.

• How can we find equivalent fractions? Again, practice with the table has prepared students for this next step.

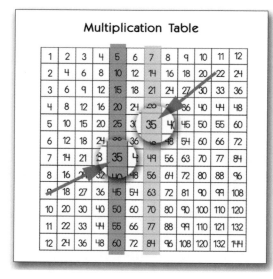

Multiplication Table

• Direct students to begin with 1/7 and find its equivalent fraction when the denominator is 35. As students follow along, cover the row beginning with 1 with one of the color strips. Use the other strip to cover the row beginning with 7. Equivalent fractions for 1/7 are now highlighted on the table.

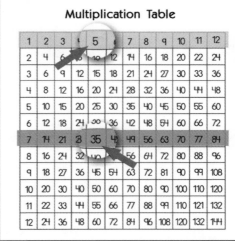

Multiplication Table

• Go to the denominator row beginning with the 7. Move your pointer to the right until you come to 35. Stop on the space with 35, then move up this column until you reach its numerator in the top highlighted row. That brings you to the 5.

• Explain that you've now used the chart to discover that 1/7 = 5/35.

• Now that you've modeled the process for finding the first equivalent fraction, ask a student to come up to the overhead and go through the same process for 2/5. She should place the acetate on the 2s row and on the 5s row.

• Let her move a finger across the 5s row until she comes to 35, and then move her finger up to the 2s row to find the equivalent numerator. She should come up with 14. So her equivalent fraction for 2/5 is 14/35.

You now have two like fractions that can be added: 5/35 + 14/35 = 19/35. Now students know how to use the chart to add unlike fractions!

You can use the same process, of course, to subtract unlike fractions. Explain to students that they must first identify the least common denominator for the fractions, then convert them to like fractions that are equivalent.

Be sure to include plenty of practice using the Multiplication Table to add and subtract unlike fractions in your Warm-Ups and Cool-Downs. It's practice that makes the process permanent!

Multiplication Table

In my presentations, when I explain the table method for working with fractions, I'm frequently asked, "When do students solve the problem mathematically, without using the table?"

The point of this approach is to ground students in the concrete when they're initially learning to work with unlike fractions; the visual process strengthens long-term understanding. Once students are comfortable using the table and have internalized the process, explain that now you're going to show them how to find common denominators and equivalent fractions mathematically. You're starting with the concrete and moving to the abstract. Most students will make a smooth transition.

Divisibility Rules

Patterns, color, and a fun introductory activity help to intrigue students with the concept of divisibility rules.

Preparation

If you haven't already done so for previous activities, make one copy of the reproducible for each student, and one overhead transparency. Cut acetate strips from report covers, making each strip the same width and length as the columns in the Multiplication Table.

MATERIALS

- Reproducible #7 (Multiplication Table)
- Acetate report covers

Directions

When you're ready to introduce divisibility rules to your students, you might do so with this motivational activity. I haven't been able to track down the source, but it makes a wonderful introduction to this concept.

To begin, tell students this is totally individual; they are not to talk or share. Then give them these directions:

- Write down a number between 1 and 9.

- Multiply that number by 9.

- Find the sum of the digits in your product.

- Take that sum and subtract 5.

- Now we're going to get away from numbers. Picture the alphabet, with *A* being 1, *B* being 2, and so on. Find the letter that corresponds to your number.

- Write down that letter.

- Now, don't think too hard, just write down quickly whatever country comes to mind that begins with that letter.

- Take the last letter of the country, and think of an animal that begins with that letter. Write down the name of the animal.

• Now that you've got that, look at the final letter of your animal's name and write a color that begins with that letter.

As soon as students have finished writing, you should say, "You know, I've never seen an orange kangaroo in Denmark."

Students will be amazed that you made that statement, because many, if not most, will have written Denmark, kangaroo, and orange as answers to your prompts.

The reason this works has to do with the pattern created by the first nine multiples of 9. For these multiples of 9, the sum of the digits is 9. So no matter what number a student starts with, as soon as he reaches the third step, his response is 9. What students don't know is that *everyone* has 9 at that point.

From there, it's a matter of probability. Most people name Denmark as the first country that comes to mind that begins with *D*. Of those who've written "Denmark," most will say "kangaroo"; a few may say "koala." If they're working with the *o* from "kangaroo" and are asked for a color, most people come up with orange. And that's how you get to the orange kangaroo in Denmark—and engage students.

Once you have their attention, you'll want to introduce divisibility rules by once again using the Multiplication Table and those colored strips of acetate. The basic principle throughout is that you place a strip of color on one set of multiples at a time. Then you encourage the class to study the pattern and hypothesize a divisibility rule for that number.

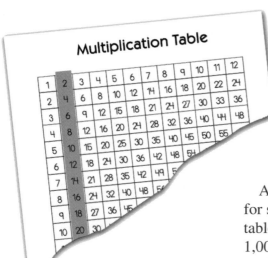

As an example, let's take a look at how you might introduce the divisibility rule for 2. We can figure out the divisibility rule for 2 by looking at the multiples of 2.

I love to do things through patterns. When I put the colored strip over the multiples of 2, I ask students what they notice about the pattern of those numbers. Then I invite someone to come up with a hypothesis about the divisibility rule for 2. It's exciting to see students suggest that a number is divisible by 2 if it's an even number.

At this point, I've found that it's both important and empowering for students to work with larger numbers that do not appear on the table. Try writing a 4-digit number on the overhead. Let's use 1,002 as your test number. (You can use this same number for testing later rules.) The number 1,002 is even, so when you ask, students should indicate it's divisible by 2. Students not familiar with large numbers might want to prove this with their calculators.

It's divisible by:

2　if the number is even.

3　if the sum of digits of the number is a multiple of 3.

4　if the number formed by the final two digits of the number is a multiple of 4.

5　if the number ends in 0 or 5.

6　if the number is a multiple of both 2 and 3.

8　if the number formed by the final 3 digits is a multiple of 8.

9　if the sum of the digits of the number is a multiple of 9.

10　if the number ends in zero.

11　if it's a 2-digit number and the 2 digits are the same; or if it's a 3-digit number and the middle number is the sum of the first and last digits (e.g., 121 or 473). When the sum of the first and last digits of a 3-digit number is a 2-digit number, subtract the middle digit from the two-digit sum. If the difference is exactly 11, the entire number is divisible by 11 (e.g., 759).

12　if it's a multiple of both 3 and 4.

Divisibility rules for 7 are complicated, and I don't usually recommend getting into them with young children.

You can repeat this discovery process as you introduce more divisibility rules. I suggest working with the rules for one number at a time, in this order: 2s, 5s, 10s, 3s, 6s, 9s, 11s, 4s, 12s, 8s.

When you get to those last 4 numbers, you'll need to go beyond the range of the numbers in the Multiplication Table. Remember the T-Chart strategy we used in Flexible Fred (page 37)? It's helpful to use this concept-attainment approach again when exploring divisibility rules for 11, 4, 12, and 8, because these go beyond the range of numbers in the multiplication chart.

Multiples of 4	
Examples	Not Examples
3,344	1,017
9,324	1,002
1,020	2,013
3,408	2,125
6,016	7,227
44	57
540	622

Once you've drawn a large T on the overhead, go to the left-hand side and write a list of numbers that are multiples of the number being studied. On the right-hand side, list ones that are *not* multiples of that number. When you're studying divisibility by 2, the left-hand list would include all even numbers, and the right-hand list would include only odd numbers. When you're introducing rules for 4, it would look something like the list on the right.

I've found that it's helpful to introduce divisibility rules gradually over a period of time, and to reinforce them frequently in Warm-Ups and Cool-Downs. In a Warm-Up, you need to be able to ask, "Is this number divisible by 2? Is it divisible by 3?" Encourage students to use their own acetate strips and copies of the Multiplication Table to prove their answers.

If you introduced these concepts and left them, they'd be gone. But if you introduce them and then provide plenty of opportunities for students to practice, you'll be providing one more opportunity for them to build that essential sense of number.

Chapter 4

MASTERING MATH FACTS ALL YEAR LONG

There's a wonderful story about a tourist visiting New York City for the first time. He approaches a police officer and asks, "Excuse me, sir, how do I get to Carnegie Hall?" The police officer responds, "Practice, practice, practice!"

Practice makes permanent. We want students to be fluent with math facts so they can use those facts in everyday problem solving. How do we help students develop fluency (also called automaticity) with math facts? We must find ways to help students to follow the officer's advice: "Practice, practice, practice!"

I believe that math facts are to math what sight words are to reading. You can't be proficient in math without having the math facts down cold, just as you can't be a good reader if you haven't mastered sight words. So students need lots of practice, and they need the right kind of practice.

Some students can go through a page of work sheet examples and never remember what they've seen or done. That's not good! We must find ways for them to become actively engaged when learning and practicing math facts, so they actually retain those facts. That's where games come in.

Because games invite a high level of engagement, I believe they offer a more effective option for building fluency than traditional work sheets. That's the reason this chapter focuses entirely on games. However, it's important to note that to gain the automaticity we're all striving for, students must play these games repeatedly all year long, not just now and then.

They must practice, practice, practice!

When students are learning and practicing math facts through games and activities, I also believe it's important to match those games and activities to student readiness and skill levels. For this reason, I begin this chapter with ways to create leveled packets of cards as well as leveled dice. Using these tools, students can advance incrementally. They might begin with a packet of cards that reinforces math facts through sums to 10. This means they're working with a leveled deck that contains numbers no greater than 5.

When they can go through that set of facts at a rate of about 25 a minute, they move on to the next leveled deck: sums to 12. Only when they reach a target speed of 25 a minute at sums to 20 do they begin using random cards.

By using polyhedral dice, we can also level dice for students to use in their games. What I've found especially helpful with this whole approach is that students are all using the same types of materials and playing the same games. They are simply working at different levels of complexity or mastery.

Leveled Card Packets

Although some students quickly learn math facts, most need active, ongoing practice. I love to use games for this, because games actively engage students in meaningful practice. I love games that use playing cards, since cards are so readily available. And what I *really* love is being able to tell students that "you don't need a full deck to play these games!"

You do need to prepare the cards, though. These directions explain how.

Directions

Begin by collecting playing cards. Keep them in a box or a large bag and don't worry about trying to keep decks intact. Remove the face cards and aces, so what you have left is a large container of playing cards with random numbers.

MATERIALS

- Several decks of playing cards
- Quart-size resealable plastic bags

These cards will be used by students who already know their addition and multiplication math facts but are not yet fluent in the use of those facts. Working with the cards will help them build up their speed to reach fluency—a combination of accuracy plus speed.

Next, create leveled packets of cards. In differentiation, this is a form of tiering to match students to their appropriate skill levels. Pre-leveled packets support students in building incremental fluency by helping them to master one level at a time.

The process for setting up these packets is easy. It goes like this:

• Separate the cards into piles of 2s, 3s, 4s, and so on, through the 10s.

• Make up packets—or have a trusted student do this—of pre-leveled groups of cards, dropping each batch in a plastic bag and labeling the bag. For example, a bag labeled "Sums to 6" would include 2s and 3s. One labeled "Sums to 8" would include 2s, 3s, and 4s. Continue grouping cards (or having students do it) for each even number up through "Sums to 20," for which the highest card included would be the 10.

• Once students have mastered all the fact families through "Sums to 20," they're ready for random practice. Create one more packet, labeled "Random Sums," and fill it with totally random cards (still excluding the face cards and aces). Have students use this packet to practice mixed addition facts.

• Also create separate leveled batches of cards for multiplication facts. In this case, create packets with random cards, but on the outside label each bag: X 2, or X 3, or X 4, up to X 12. (With the multiplication sets, it doesn't matter which cards go in each bag; the difference is in the label, which tells a student which family of multiplication facts he's to practice with that packet.)

• Instruct each student to pull one card at a time from the multiplication packet he's working with, and to multiply the number on that card by the number on the packet's label. A student beginning to learn multiplication facts might work with 2 as a multiple. Another student or group might play a game that involves practicing multiples of 9.

• As with the addition process, students work their way through the multiplication fact families. Once they reach the level of 12s, and are able to do about 25 a minute, you need to provide one more packet. This one is labeled "Mixed Multiplication Facts." A student using this packet pulls 2 cards from the packet at a time and multiplies the number on one card by the number on the other.

• Once you've identified the level that's appropriate for each student or small group, have the student(s) take that packet of cards and practice as described above and by playing games such as the ones in this chapter. Students are on their way to building automaticity!

CHANGE of PACE

PRACTICE THE BASICS

For quick practice of a basic geometry concept, have students line up as a line segment. Instruct one student to become "point A," at the beginning of the line, and another student (one who typically bolts to the front) to become "point B," at the end. You might also add a midpoint. (This is a great math and behavioral management technique.)

Captive Dice

Captive dice capture students' attention and give them a chance to practice math facts in a game situation. The closed containers also keep students from losing the dice.

Directions

To create your own sets of captive dice, put two dice or number cubes in each container, then close or even superglue each lid in place so the dice can't escape.

When you want to "bump it up a notch," you can capture polyhedra. Polyhedra are solids that have flat sides. These flat sides or faces are called polygons. There are only five regular polyhedra, all of which you can buy in roughly the same form as regular dice. Use different variations to level dice games for your students; the greater the number of faces on the polyhedron, the greater the challenge.

To increase the difficulty level for advanced students beyond hexahedra (number cubes), put together captive combinations of octahedra, dodecahedra, or icosahedra (see page 85). You now have leveled containers of dice that, much like the cards, can be used in skill-building games throughout the year.

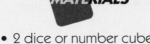

MATERIALS

- 2 dice or number cubes per set
- Small plastic containers with transparent bottoms

Teacher Tip

Grocery stores, discount stores, and yard sales are all great sources for these plastic containers.

Building a Tetrahedron & a Hexahedron

To introduce students to the polyhedra they'll be using throughout the year, try this motivational lesson, which involves making models with gumdrops and toothpicks.

Directions

Follow these steps:

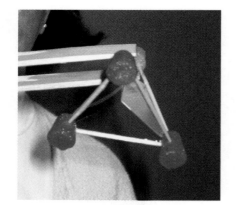

• Mix the dishwashing liquid and water together in the pail until you get a good strong soap mixture and can blow bubbles through a bubble wand.

• Piece together the gumdrops and toothpicks to create a regular tetrahedron.

MATERIALS

- 1 cup Ultra Dawn dishwashing liquid
- 2 quarts water
- 1 gallon-size plastic pail or other container
- 1 large bag gumdrops
- 1 or more boxes toothpicks (regular or dinner size)

• Dip the tetrahedron in the soap mixture. (You may have to repeat this step a couple of times.) When you remove it from the soapy water, the surface tension along with the position of the vertices inverts the three faces of a triangle inward, within each face of the tetrahedron. No matter which way you look at it, you can see the edges, faces, and sides. Voilà: a tetrahedron!

• Invite each student to try this for himself.

• To make a cube, repeat exactly the same process, but instead of constructing a tetrahedron, use the toothpicks and gumdrops to build a cube, also known as a regular hexahedron. When you dip the cube into the soapy water, students will discover that as they look at each face, they'll see four inverted sides and a cube at the center of the hexahedron. This is known as a tesseract.

• Again give students a chance to do this on their own.

Warning: this science and math activity can lead to "soaperior" learning!

Perhaps you have heard about "Platonic Solids." In ancient Greece, Plato discovered that there are five regular solids. A regular solid is one with congruent edges, faces, and angles. Congruent means each would fit exactly over the other; they have exactly the same shape and size.

These are the five regular poly-hedra; all can be purchased at most teacher stores or from math product distributors such as Lakeshore, ETA/Cuisenaire, and Learning Resources.

• A **tetrahedron** is a triangle-based pyramid; its 4 faces are equilateral triangles. There are 3 numbers on each face of a commercial tetrahedron. For this reason, I don't use these a great deal in regular classroom games.

• A **hexahedron** is a cube. Its 6 faces are squares. If you create captive dice with 2 cubes, a student can practice sums to 12 and products to 36.

• An **octahedron** has 8 faces, all equilateral triangles. Create captive dice with 2 octahedra, and a student can practice sums to 16 and products to 64.

• A **dodecahedron** has 12 faces, all pentagons. Place 2 dodecahedra in a container, and students can practice sums to 24 and products to 144.

• An **icosahedron** has 20 faces, all equilateral triangles. With 2 icosahedra in a container, students can work with sums to 40 and products to 400.

The Facts of Life

I explain to students that the facts of life are that they must know their math facts! You can level this game and assign it to groups of two or more. To differentiate for students' readiness or skill levels, assign them appropriate packets of leveled cards. To increase the complexity, have students treat the red cards as negative integers and the black cards as positive ones. For all groups, the goal remains the same: to determine the greatest product.

Preparation

If you haven't already done so for previous activities, make one copy of the reproducible for each student and cut strips from the report covers. Each strip should be the same width and length as the columns in the table.

MATERIALS

- Reproducible #7 (Multiplication Table)
- Colored acetate report covers
- Random or leveled playing cards

Directions

- One student deals all the cards, face down, to all the players.

- Each player turns over two cards and multiplies the numbers on those two cards.

- The student with the greatest product (not the one who gets it first) is the winner and gets all the cards.

- Students use the acetate strips on the Multiplication Table to double-check their answers.

- Repeat until all the cards have been used.

In the event of a tie, cards from the "tie round" are left on the table. Students complete another round of multiplying. Winner takes all!

Variations

Variation 1: Use the same approach, but instead of multiplying the numbers on the cards, find the sums of the numbers or find the differences between them.

Variation 2: To increase the challenge, turn over three cards at a time.

Variation 3: Use captive dice (see page 83) in place of cards.

First with the Facts

A simple card game for groups of two or more builds automaticity with multiplication facts. It's easy to level this game appropriately for each of your students.

Preparation

If you haven't already done so for previous activities, make one copy of the reproducible for each student and cut strips from the report covers. Each strip should be the same width and length as the columns in the table.

MATERIALS

- Reproducible #7 (Multiplication Table)
- Colored acetate report covers
- Leveled or random (depending on group's skill level) playing cards

Directions

• One student deals out a stack of leveled or random playing cards, face down, distributing them evenly among the players.

• The dealer turns over one of his cards. That card becomes the "constant" factor in the game.

• The student to the dealer's left turns over a card from her stack. That card is the multiplier. All students multiply this number by the "constant" factor in the center.

• The first student to name the correct product gets the multiplier card.

• Students use the acetate strips on the Multiplication Table to double-check their answers.

• The play continues around the table, with each student turning over a multiplier card. Throughout the game, players always use the same "constant."

• Each student continues to play until her pile is completely used.

Variations

Variation 1: Students can also continue playing the game through rounds. When all cards have been used, one round is over. The winner of that round is the student with the most cards. The dealer then reshuffles the cards and deals them again. Students play another round. Again, the student with the most cards wins the round. When students are finished playing the game, the student who has won the most rounds wins for that session.

Variation 2: Students can also play this same game to practice addition facts instead of multiplication facts. In this case, of course, they use copies of Reproducible #6 (Addition Table) to double-check their answers.

The constant is 8. The multiplier is 6.

A SOURCE FOR CARDS AND DICE

Casinos can be a great source for free playing cards! They use each deck of cards on a very limited basis, then they punch a hole in each card and throw the whole lot away. Try writing to a casino, using your school letterhead, and asking if you can have some of the cards they'd otherwise discard. (The cards typically have the casino's logo on the back, so you'll need to explain to parents what you're doing.)

Casinos also use dotted dice on a limited basis. As with the cards, they use the dice briefly, then punch holes in them and discard them. Use your school stationery to request used dice, too!

Facts on the Brain

The object of this game for three students is to practice finding products and missing factors. One player is the product finder, while the other two are factor finders.

Preparation

If you haven't already done so for previous activities, make one copy of the reproducible for each student and cut strips from the report covers. Each strip should be the same width and length as the columns in the table.

MATERIALS

- Reproducible #7 (Multiplication Table)
- Colored acetate report covers
- Playing cards

Directions

• The product finder deals each factor finder a pile of leveled or random cards, placing them face down.

• The product finder counts aloud, "1, 2, 3!"

• On "3," each factor finder lifts the top card from his pile and places it on his head.

• The product finder announces the product of the two numbers.

• Knowing the product, each factor finder looks at his partner's number to determine the missing factor.

• The first student to announce the correct missing factor wins that round and gets both cards.

• Students use the acetate strips on the Multiplication Table to double-check their answers.

• The students repeat this process with their remaining cards until they've used all their cards.

• When they've used up their cards, students rotate roles, shuffle and deal the cards, then begin again.

Object Math

This is a game for individual students, though it can also be used for pairs or small groups. It uses the "guess and check" problem solving strategy while incorporating math fact practice. You can level the game by including cards only up to certain numbers or by asking students to use only addition and subtraction functions.

Preparation

If you haven't already done so for previous activities, make one copy of each reproducible for each student and cut strips from the report covers. Each strip should be the same width and length as the columns in the Addition Table.

Directions

• Working from a full deck or a leveled set of cards, a student turns over three cards and places them side by side.

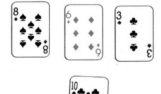

• He then turns over one more card and places it below the first three. This last card is the object card.

• The student must use each of the three cards once to make an equation that equals the object card. Any or all of the arithmetic operations can be used, and the order of the cards doesn't matter.

• If an individual student is playing this game, she simply tries to find at least one valid equation or continues to find as many as possible within a given time limit. Then she sets those cards aside, turns over another four cards, and repeats the process.

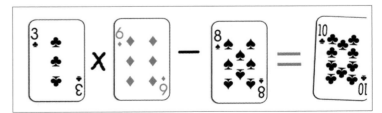

• With partners or a small cooperative group, students collaborate to find a valid solution.

• No matter what the group size, students use the acetate strips on the tables to double-check their answers.

Smath

Similar to the playground game Spud, this is an ideal, active way to practice math facts. The teacher is in control of the difficulty level of each problem and can match expectations to students' skill levels. The class or small group can play this game in the gym or on the playground.

Preparation

Take a class list and pre-assign a number to each student. Make sure each student knows his number.

MATERIALS

- Playground ball or any soft ball

Directions

- Students stand in a wide circle around the teacher.

- The teacher counts "1, 2, 3," then shouts out part of a math fact (e.g., "5 + 6!") and throws the ball into the air.

- The student whose number is the answer must run into the center and catch the ball before it hits the ground, then toss it back to the teacher.

- The game continues with the teacher beginning another number sentence and again throwing the ball into the air.

"DISCOVERING" MATH SKILLS & CONCEPTS

T his chapter includes a variety of ways to strengthen students' conceptual knowledge of specific math topics, as well as ideas for reinforcing important mathematical principles. The goal of this chapter is to provide alternative pathways to learning.

Here you'll find ideas to help students internalize the concept of place value, a fresh approach to teaching shapes and measurement, and a new and logical way to show students how to factor numbers.

I've included a hands-on method to introduce the concept of square numbers and square roots to students who need a concrete foundation, and a patterned activity to make subtraction fun. Other strategies, including those involving estimation and statistics, are well worth using in your classroom on a regular basis to reinforce learning. We need to fill our instructional backpacks with as many strategies as we can.

I firmly believe that if students are not learning from the way that we teach, then we need to teach them the way that they learn. These activities are designed to do just that.

Place Value Participation

I t's vital that we strengthen students' understanding of place value. Our Hindu-Arabic numeration system is a positional system, based on groupings of 10. Students must learn that the face value of a number is determined by its place value. For example, in the number 265, the 6 does not represent a face value of 6, but rather 6 groups of 10.

This approach offers a wonderful way to actively engage students and reinforce their understanding of place value. When you introduce this activity, choose numbers that are appropriate for your group, making sure you begin with easy examples and work toward more difficult ones.

Preparation

Using a bold black marker, print one whole number on each of the index cards. Use the following whole numbers: 0, 1, 2, 3, 4, 5, 6, 7, 8, and 9. Repeat the process twice, so you end up with 3 cards for each number.

MATERIALS

• 30 index cards (4" X 6")

Directions

• Give each student one or more of the number cards you've made. (It doesn't matter if some students have one and others have more.)

• Call out a number. Let's say you call out 325.

• Students listen and examine their numbers. Anyone holding a 3, a 2, or a 5 should move to the front of the room. Since three students have each of these numbers, the goal is to be one of the first to recognize his number, move to the front of the room, face the class, and form the number 325.

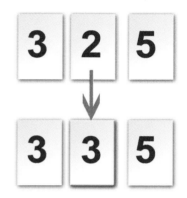

• Ask for someone to come up and make the number that's 10 more than 325. All students should examine the tens place in 325, mentally increase it by 10, and determine the value would now be 335. A student holding a 3 should come up and replace the student who's holding the 2 and currently standing in the tens place.

• Ask for someone to come up and make the number 300 more than 325. Repeat the process.

• Continue in this fashion, each time inviting students to actively represent the value of the revised number.

Variation

You can purchase sets of color-coded place value strips (see Crystal Springs Books in the resource section) designed by a math teacher named Bob Hogan. In these strips the units, tens, hundreds, thousands, ten thousands, hundred thousands, and millions are in different colors. Give one set to each student, and you can play "Place Value Participation."

In this version of the activity, all students remain at their seats or tables and use their strips to create the numbers you dictate. You can walk around and observe students in the process of creating numbers. When you ask students to add 300 to the value, every student works through the process with the strips. You can check immediately for understanding and offer help where needed.

ON THE MOVE

Every once in a while, give directions mathematically. For example, when lining up for PE, tell students to line up in a 180° angle, perpendicular to the door. Ask them to walk straight ahead 2 yards into the hall, then turn 90° to their left. Tell them to continue to the gym, and ask them to stand in a 360° circle and wait for further instructions.

Subtraction Action

Many times students will do subtraction problems in what I call the "whichever-way-works" mode. If the problem is 64 - 25, they start with the units and they say "5 take away 4 is 1." Then they go to the tens and they say, "6 take away 2 is 4. So the answer is 41." Clearly, this is a problem.

This activity helps students who understand subtraction, but have difficulty with procedural computation. It requires students to set up a subtraction problem before they can solve it, which helps them stop and think.

One of the great things about Subtraction Action is that it can be used again and again because there's more than one correct answer, and there's always a pattern in the center. Students learn that if there's no pattern in the center, they need to go back and check their math.

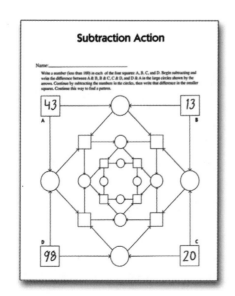

Preparation

Make one copy of the reproducible for each student and one overhead transparency.

MATERIALS

• Reproducible #8

Directions

Place the transparency on the projector and model how to complete the activity:

• Point out the outer corner squares labeled A, B, C, and D.

• Ask students to give you four numbers, each less than 100.

• Put one of these numbers in each of the outer squares.

• For example, if students give the numbers 43, 13, 20, and 98, you would record one number in each outer square, A, B, C, and D, respectively.

• Show students how they need to compute the differences between the numbers in the squares, and write their answers in the circles. (See the illustration.) They've just completed four subtraction problems, and they needed to set them up either on paper or in their heads. Setting up the problem with the greater number first encourages students to think before they compute.

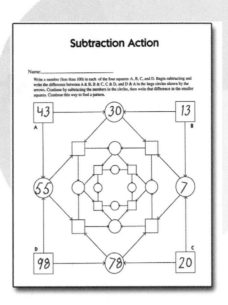

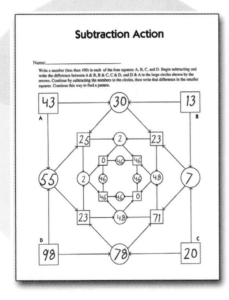

• Do four more problems by subtracting the numbers in the circles and writing their differences in the smaller squares.

• Repeat this process, following the arrows as you get closer and closer to the center.

Follow up by giving students a chance to practice on their own. Be sure they understand that they should always end up with a pattern in the center iteration. It may or may not be zeros, but there should always be a pattern. Students are fond of this activity, and I like it because it's a fun and painless way to practice subtraction.

Finding Square Numbers & Their Roots

Students sometimes have difficulty understanding square numbers and square roots of numbers. This is a way I've found to help them learn about these functions, beginning at a concrete level.

In this activity, students begin with hands-on manipulatives, progress to the pictorial, and then move on to the abstract, at which point they learn to "trust the math."

Preparation

Make one paper copy of the reproducible for each student and one overhead transparency.

MATERIALS

- Reproducible #1
- 25 one-inch tiles for each child
- 25 one-inch overhead tiles

Directions

• Place the transparency on the overhead. Note that the bottom row of the grid is labeled as the "root."

• When introducing this method for the first time, invite students to tell you the function of a tree's root. They'll tell you that the tree or plant grows from the root. Explain to students that numbers also have roots. It's the base from which they grow.

• Explain that "if we consider the root to be the bottom row on this grid, we can begin by creating a root of 2 [placing 2 tiles on the bottom row of the grid]."

• Continue, "Now, if we want to square the number 2, we do that by taking tiles and building a square." Do this by placing 2 tiles immediately above the "root" 2.

• As you demonstrate on the overhead, have the

students do the same thing with their tiles at their seats. Ask if they see the square. They've each built a 2 by 2 square and can concretely see that 2 squared is 4.

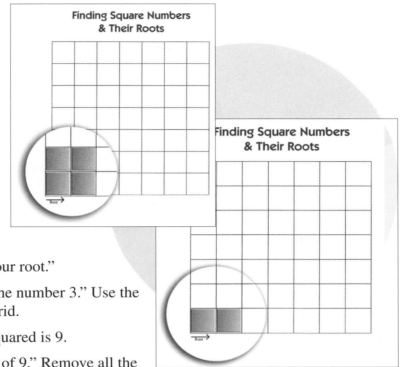

Finding Square Numbers & Their Roots

Finding Square Numbers & Their Roots

• Once they see the square, ask them what the root of 4 is. Show them how they can remove tiles, so that the only tiles remaining are the 2 in the root. The root of 4 is 2.

• Model for students again by squaring the number 3. This time place three tiles on the bottom row and say, "We now have 3 squares in our root."

• Continue by saying, "Let's square the number 3." Use the tiles to build a 3 by 3 square on the grid.

• Count the tiles and explain that 3 squared is 9.

• Say, "Now, let's go back to the root of 9." Remove all the tiles except those in the root.

• Explain that the root of 9 is 3.

• Students who still need additional concrete experience can continue using the tiles to find the square of 4, then the square root of 16. They might also continue and square 5, then find the root of 25, but eventually I tell them they have to trust the math!

Doing this exercise—letting students see square roots in a concrete, visual way—helps them store the concept in long-term memory.

After using the discovery method to introduce square numbers and their roots, take students to the pictorial level by showing them how to write these terms mathematically. You will want to show students that we write 2 squared as 2^2, and that we write the square root of 4 like this: $\sqrt{4}$. Have them practice adding the symbols, so the understanding and use become automatic.

Of course, it's best when you reinforce this by including square numbers in your Daily Number Review as part of Warm-Ups and Cool-Downs. If you were reviewing the number 3, you would now include a question about 3 squared. You might have students get out their square root grids and tiles, and ask them to build a square from the root of 3. They'll get 9. And they'll see that the square root of 9 is 3.

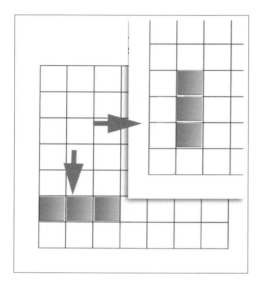

Now ask students if they can create a square anywhere on the grid using just 3 tiles. They'll discover that they can't. They can line up 3 tiles horizontally, and they can line up 3 tiles vertically, but they can't make a square. So they'll discover that 3 is rectangular. You can square 3 to get 9, which means conversely that the square root of 9 is 3. But the number 3 by itself is not a square number.

Another day, you might be reviewing the number 6, which can be squared to 36 but is not itself a square number. Once again, if students use 6 tiles to attempt to make a square anywhere on the grid, they'll find it's not possible. By experimenting with the tiles, they'll find they can create rectangles of 3 X 2, 2 X 3, 1 X 6, and 6 X 1. In this way, not only do students discover that 6 is a rectangular number, but they also begin to see the concept of area and the commutative property of multiplication.

When you introduce and reinforce concepts with a concrete, hands-on approach, it strengthens students' understanding in a powerful way.

SQUARE NUMBERS FROM ANOTHER DIRECTION

Here's another way you can reinforce the concept of square numbers and their roots. Take the Multiplication Table (reproducible #7) and put bingo chips on the numbers in a diagonal direction, beginning with the 1 in the upper left-hand corner and moving downward to the lower right-hand corner. Do you notice what happens? Do you see that you have all square numbers?

If you stop on any one of these square numbers, you can find its root by moving vertically to the top of its column. Let the students find the pattern with the table. It's one more way to reinforce the concept in students' long-term memories.

1	2	3	4	5	6	7	8	9	10	11	12
2	4	6	8	10	12	14	16	18	20	22	24
3	6	9	12	15	18	21	24	27	30	33	36
4	8	12	16	20	24	28	32	36	40	44	48
5	10	15	20	25	30	35	40	45	50	55	60
6	12	18	24	30	36	42	48	54	60	66	72
7	14	21	28	35	42	49	56	63	70	77	84
8	16	24	32	40	48	56	64	72	80	88	96
9	18	27	36	45	54	63	72	81	90	99	108
10	20	30	40	50	60	70	80	90	100	110	120
11	22	33	44	55	66	77	88	99	110	121	132
12	24	36	48	60	72	84	96	108	120	132	144

Partner Factors

The typical method that many texts use to show students how to factor numbers has left many of my students confused and perplexed. Instead, I like to teach them my own method, which I call Partner Factors. It helps students create an organized list, plus it creates a visual way to recognize and work with factors.

Directions

In my class, we'd create small partner factor charts on index cards and post them around the room, much like a number line. You can do the same thing, creating each mini-chart together with your students, sometimes during a Warm-Up or Cool-Down.

MATERIALS

• 180 index cards (3" X 5")

Let's say you're on the 16th day of the school year, and you're factoring the number 16.

• First, draw a vertical line down the center of the card, then write the target number of 16 in the upper right-hand corner, next to the vertical line.

• Next, ask questions. Begin with 1. Ask, "Is 1 a factor of 16?" They'll say yes; 1 is a factor of every number.

• Ask, "And what is its partner?" In this example, you want them to say 16. Show them how to write 1 directly across from the 16, but to the left of the vertical line.

• Ask, "Is 2 a factor of 16?" It is, so you list the 2 on the left of the line, and its partner factor, 8, to its right.

• Next ask, "Is 3 a factor?" No, it's not, so go on to 4.

• Ask, "Is 4 a factor of 16?" It is. List a 4 on each side of the vertical line.

• Explain to students that you know you now have all of the factors, because they're starting to repeat. You already have 2 and 8 as partners. Listing 8 and 2 would just be repeating what you already have.

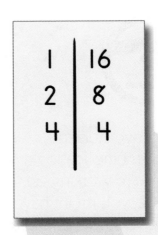

You can do the same thing with 18. Begin by writing an 18 at the top of the right-hand column, and then listing its partner, 1, to its left. Then go down the columns and list 2 with its partner, 9, and 3 with its partner, 6. That's how you teach students to factor numbers.

As we created these cards, we posted them around the room in number-line form, beginning with 1 and ending with 180. Students now had visual tools they could use to reinforce multiplication facts. They could find least common multiples and greatest common factors, and they could identify prime and composite numbers. Finally, they could use the cards when working with fractions.

You can create these cards to use throughout the year, or simply show students this easy way to factor numbers. What you're doing, of course, is playing with numbers, helping students to look for patterns and build number sense.

WORK WITH ARRAYS

Have students create multiplication arrays. If you're working on factors of 12, you might have twelve students move about to physically form:
• two rows of six, then six rows of two
• three rows of four, then four rows of three
• one row of twelve, then one column of twelve.

Guesstimate/Bestimate

This weekly activity is a fun and effective way for students to build estimation skills.

Preparation

Before class, pour some of the manipulatives into one of the containers. Leave the other one empty for the time being. Make an enlarged copy of the reproducible, write your students' names or assigned numbers in the left-hand column, and then post the resulting chart on the class bulletin board.

MATERIALS

- Small manipulatives or materials that can be counted (e.g., Cheerios, seashells, buttons, beans)
- 2 clear plastic containers of the same size
- Reproducible #9

Directions

When you introduce Guesstimate/Bestimate, you might explain that this will be a weekly activity. It's a good idea to begin it on a Monday, so students have a whole week to fine tune their answers. The process for completing the activity will be the same each week, but the material in the estimating container will change.

How does the activity work?

• Show the students the container of Cheerios (or other manipulatives).

• Ask the class what they need to do when there are too many to count. They'll tell you to estimate the quantity.

• Give everyone a chance to guess how many Cheerios are in the container. Tell students that by Wednesday, each of them should write her "guesstimate" in the column next to her name. Remind them that a "guesstimate" is a very rough estimate.

• On Wednesday, check to make sure everyone has written down a "guesstimate."

• On Friday, give students a chance to change their answers and give "bestimates," based on new, specific information.

• What do you do to improve "guesstimates"? Take the container with the Cheerios, along with the identical container that's been

Teacher Tip

Depending on the class, you might want to assign each student a number and have everyone write his estimate next to his number. Keeping names confidential avoids any embarrassment when estimates are way off the mark.

Guesstimate / Bestimate

This Week's Estimation Question: _How many Cheerios in the container?_

Student/Group	Guesstimate	Bestimate
A	200	220
B	60	125
C	250	220
D	400	300
E	500	400

Teacher Tip

If you have a large class and too little time (don't we all?), you can have groups of students estimate. That gives you a smaller sample of data, so you can do the statistical analysis more quickly.

empty up to this point. Pour approximately half of the Cheerios (in this case) into the second container.

• Assign one student to count how many Cheerios are in one of the containers.

• Ask the student to share the answer with the class.

• Remind students that they now know how many are in half of the original total. Ask students if they now want to give a "bestimate," based on new mathematical information.

• Write their new estimates in the "bestimate" column on the chart. (Students can change their own "guesstimates," but it takes more time.)

• Work together to determine this week's winner—the student whose "bestimate" is closest to the right answer. That child claims the title of "Bestimator of the Week," and the reward is to bring in the estimating material for next week's activity. (Since students sometimes forget or don't have access to materials, it's a good idea to always have plenty of options in your teacher drawer.)

Of course, if the student counts 100 Cheerios in that "half batch," you're hoping that every child changes his "guesstimate" to a "bestimate" of 200 when given the chance. You'll learn a great deal about students' number sense through this activity.

After a few weeks of repeating this activity, I find that student estimation skills improve. And as long as you're gathering all that data, you can go one step further and have the class work together to analyze the information. That's where the next activity comes in.

Putting Statistics on the Line

This interactive learning activity, repeated on a weekly basis, builds student understanding of statistical terms and concepts.

Preparation

Make a transparency of the reproducible. Then string a length of clothesline or yarn across part of the classroom.

Directions

Let's say that today we want to work with the data from our "guesstimate/bestimate" activity. To simplify, let's also assume that students worked in groups this week. What are the steps to Putting Statistics on the Line?

MATERIALS

- Reproducible #10
- A length of clothesline or yarn, long enough to stretch across part of the room
- Data to be analyzed
- Post-it notes

• Give each group one Post-it note and a dark felt-tipped marker. Have each group write their final "bestimate" on a Post-it note. (Tell them to print large numerals.)

• Ask the groups to call out the "bestimates" on their Post-its. List the "bestimates" on the overhead or write them on the board.

• Identify the highest and the lowest "bestimates." Attach the Post-it with the lowest value on the left end of the clothesline. Attach the Post-it with the highest value on the far right end of the clothesline. If there are duplicate values, create a vertical column by sticking one Post-it to the bottom of another. (You might need to use tape.)

• Attach the values between the highest and lowest, correctly spacing them on the clothesline.

• Let's say that this week Group A guessed that there were 220 Cheerios, Group B guessed 125, Group C also guessed 220, Group D guessed 300, and Group E guessed 400. The line would look something like the photo at the right.

• Ask students to help you record the data on this week's chart, starting with the sample size. Students should tell you there are 5 values, so the sample size is 5.

• Compute the range. Ask students to find the difference between 400 and 125. That would be 275.

• Is there a mode? Invite someone to define the mode, then explain that the value of 220 appears twice in our data. It's the value that appears most frequently, so it's the mode.

• Ask, "What's another word for mean?" You hope students will know it's the average value. Ask them to compute on paper or to use a calculator. Remind them how they averaged numbers using the Hundred Chart and how they also computed the average. Ask, "What were the steps? First you add the values, then you divide by the number of addends. What's the mean for this set of data? 253." Write that on the chart.

• Finally, determine the median. The middle value in this set is 220. That's the median.

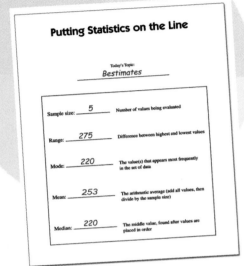

• Now that the information is listed on the chart, analyze it. Ask students if they can draw any conclusions from this data.

The beauty of putting these numbers on a clothesline is that it lets students see the mode and the range of values. As you collect and analyze data on a regular basis, students will not only come to understand the meaning of these statistical terms but will also learn how to analyze the data and how to determine what, if any, conclusions can be drawn. The concepts become so much clearer when students can see the numbers and put statistics on the line!

Measurement—What Shape Are You?

Our human bodies are amazingly proportional. This activity takes advantage of that fact, reinforcing geometry concepts by letting each student determine whether he's a square or a rectangle.

Directions

This is a hands-on geometry lesson in which students work with partners or in small groups. Let's say they're in pairs. Ask students, "What shape are you? Are you a square or a rectangle?" If Ryan and Matt are an example of one pair, proceed as follows:

MATERIALS

• Yarn

• Have Ryan stand up straight. Ask Matt to measure his (Ryan's) height with yarn.

• Have Matt cut a piece of yarn that represents Ryan's height.

• Next, have Ryan stretch his arms out at his sides. Ask Matt to cut another piece of yarn that fits Ryan's arm span from middle finger to middle finger.

• Have Matt and Ryan take the two pieces of yarn and compare them.

• If the length and width are exactly the same, Ryan is a square. If one is longer than the other, he's a rectangle.

• Have Matt and Ryan trade roles so Ryan can measure Matt.

• Gather the data for all of the students and graph it to show how many squares and how many rectangles are in the class.

What's Your Circumference?

This activity offers a great way to teach about circles. Take the students outside to the playground, but bring along a skein of yarn!

Directions

Measure each student's height with the yarn and cut the length of yarn to exactly match his height. Explain that the yarn represents that child's personal diameter.

• Yarn

We know that the radius of a circle is half of its diameter. So now, ask the child to take the yarn and fold it in half to represent his personal radius.

Explain that the student should take one end of the folded yarn and anchor it with a rock. He should then use the folded length of yarn as a compass and draw a circle with it. If you're working on a dirt playground, he can use a stick to mark the circle in the dirt. If the playground is covered with asphalt, he can make the circle with a piece of chalk.

When the circle is completed, the child should be able to lie down and fit perfectly inside his personal circle, with his head at one point on the circle and his feet directly opposite. He's now representing his own diameter.

RECOMMENDED RESOURCES
Books for Math Teachers

Chapin, Suzanne H. and Art Johnson. *Math Matters: Understanding the Math You Teach.* (Grades K-6.) Sausalito, CA: Math Solutions Publications, 2000.

Crooks, Lisa. *Connecting Math with Literature: Using Children's Literature as a Springboard for Teaching Math Concepts.* (Grades 3-6.) Huntington Beach, CA: Creative Teaching Press, 2002.

Janes, Nancy Segal. *Exploring Math with Polyhedra Dice: Skill Building Activities in Problem Solving.* Vernon Hills, IL: Learning Resources, 2001.

Ma, Liping. *Knowing and Teaching Elementary Mathematics.* Mahwah, NJ: Lawrence Erlbaum Associates,1999.

Parker, Thomas H. and Scott J. Baldridge. *Elementary Mathematics for Teachers.* Bloomington, IN: Sefton-Ash Publishing, 2003.

Principles and Standards for School Mathematics. Reston, VA: National Council of Teachers of Mathematics, 2000.

Children's Books

AREA AND PERIMETER
- Burns, Marilyn. *Spaghetti and Meatballs for All!* New York, NY: Scholastic, 1997.

ESTIMATION
- Murphy, Stuart. *Betcha!* Mathstart Series. New York, NY: HarperCollins, 1997.

FRACTIONS
- McMillan, Bruce. *Eating Fractions.* New York, NY: Scholastic, 1991.

MONEY
- Viorst, Judith. *Alexander, Who Used to Be Rich Last Sunday.* New York, NY: Aladdin, 1978.

NON-STANDARD MEASURE

- Myller, Rolf. *How Big Is a Foot?* New York, NY: Dell, 1990.

NUMERATION

- Alda, Arlene. *1 2 3 What Do You See?* Berkeley, CA: Tricycle Press, 1998.

- Friedman, Aileen. *The King's Commissioners.* New York, NY: Scholastic, 1994.

- Mathis, Sharon Bell. *The Hundred Penny Box.* New York, NY: Puffin Books, 1975.

- Murphy, Stuart. *Hundred Days of Cool.* New York, NY: HarperCollins, 2003.

- Tang, Greg. *The Best of Times.* New York, NY: Scholastic, 2002.

POETRY

- Pappas, Theoni. *Math Talk: Mathematical Ideas in Poems for Two Voices.* San Carlos, CA: Wide World Publishing, 1991.

POLYGONS

- Burns, Marilyn. *The Greedy Triangle.* New York, NY: Scholastic, 1995.

- Hoban, Tana. *Shapes, Shapes, Shapes.* New York, NY: HarperCollins, 1986.

- Tompert, Ann. *Grandfather Tang's Story.* New York, NY: Crown Publishers, 1990.

PROBLEM-SOLVING STRATEGIES

- Cousins, Lucy. *Maisy's Mix-and-Match Mousewear.* Cambridge, MA: Candlewick Press, 1999.

- Schwartz, David M. *How Much Is a Million?* New York, NY: Lothrop, Lee & Shepard, 1985.

- Scieszka, Jon and Lane Smith. *Math Curse.* New York, NY: Viking, 1995.

- Tang, Greg. *Math-Terpieces: The Art of Problem Solving.* New York, NY: Scholastic, 2003.

RATIO & PROPORTION

- Briggs, Raymond. *Jim and the Beanstalk.* New York, NY: Sandcastle Books, 1970.

- Schwartz, David M. *If You Hopped Like a Frog.* New York, NY: Scholastic, 1999.

REFERENCE BOOKS

- de Klerk, Judith. *Illustrated Math Dictionary.* Parsippany, NJ: GoodYear Books, 1999.

- Monroe, Eula Ewing. *Math Dictionary for Young People.* Morristown, NJ: Silver Burdett, 1998.

- Pappas, Theoni. *Math for Kids & Other People Too!* San Carlos, CA: Wide World Publishing, 1997.

- Schwartz, David M. *G Is for Googol.* Berkeley, CA: Tricycle Press, 1998.

- Wong, Harry K. and Rosemary T. Wong. *The First Days of School.* Mountain View, CA: Harry Wong Publications, 1998.

TESSELLATIONS

- Friedman, Aileen. *A Cloak for the Dreamer.* New York, NY: Scholastic, 1995.

Product Sources & More

Boxcars and One-Eyed Jacks
www.boxcarsandoneeyedjacks.com
Alberta, Canada: 1-780-440-6284

Crystal Springs Books
www.crystalsprings.com
1-800-321-0401

Delta Education
www.deltaeducation.com
1-800-258-1302

ETA/Cuisenaire
www.etacuisenaire.com
1-800-445-5985

Lakeshore Learning Materials
www.lakeshorelearning.com
1-800-778-4456

Learning Resources
www.learningresources.com
1-800-333-8281

National Council of Teachers of Mathematics
www.nctm.org
1-800-235-7566

U.S. Toy Company
www.ustoy.com
1-800-832-0224

PHOTOS FOR CLASSROOM USE

PHOTOS FOR CLASSROOM USE

Finding Square Numbers
& Their Roots

Root →

A Geometry Scavenger Hunt

Name: _____

Can you find each item in the room, or find someone who is wearing:

1. triangles? _____

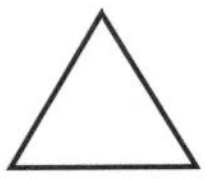

2. squares? _____

3. rectangles? _____

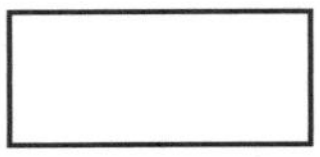

4. circles? _____

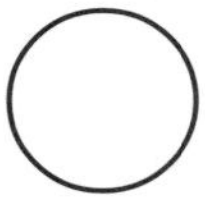

5. circles & squares? _____

6. hexagons? _____

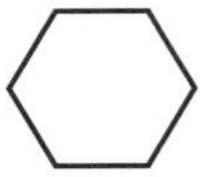

7. a tessellating pattern? _____ 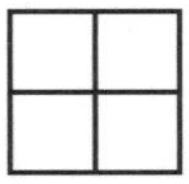
(a repeating pattern with no gaps)

8. 3 x 2? _____

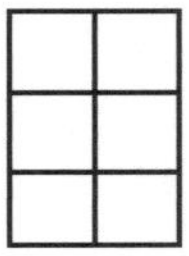

REPRODUCIBLE #2

Daily Number Review

Name: _____

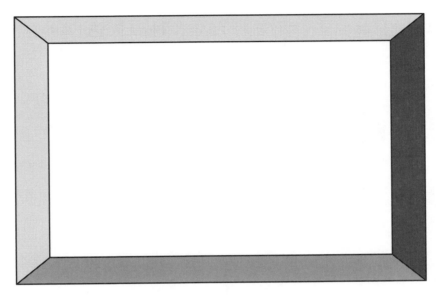

Number Properties & Terms

Circle the correct properties & terms for today's number.

Even or Odd

Prime or Composite

Palindrome

Integer

Counting/Natural Number

Whole Number

Square

Picture Puzzles

1. STAND
   MIS

2. __0__
   B.S.
   M.A.
   Ph.D.

3. BOOK
   DUE

4. WALKING
   EGGSHELLS

5. SKATING
   iiiiiiii

6. WALKING
   AIR

7. THEHCATAT

8. __RIGHT__
   TIME

9. SIDE/SIDE

10. __MAN__
    BOARD

11. MYSOMETHINGPOCKET

12. AWONCEHILE

Hundred Chart

1	2	3	4	5	6	7	8	9	10
11	12	13	14	15	16	17	18	19	20
21	22	23	24	25	26	27	28	29	30
31	32	33	34	35	36	37	38	39	40
41	42	43	44	45	46	47	48	49	50
51	52	53	54	55	56	57	58	59	60
61	62	63	64	65	66	67	68	69	70
71	72	73	74	75	76	77	78	79	80
81	82	83	84	85	86	87	88	89	90
91	92	93	94	95	96	97	98	99	100

Addition Table

+	0	1	2	3	4	5	6	7	8	9	10	11	12
0	0	1	2	3	4	5	6	7	8	9	10	11	12
1	1	2	3	4	5	6	7	8	9	10	11	12	13
2	2	3	4	5	6	7	8	9	10	11	12	13	14
3	3	4	5	6	7	8	9	10	11	12	13	14	15
4	4	5	6	7	8	9	10	11	12	13	14	15	16
5	5	6	7	8	9	10	11	12	13	14	15	16	17
6	6	7	8	9	10	11	12	13	14	15	16	17	18
7	7	8	9	10	11	12	13	14	15	16	17	18	19
8	8	9	10	11	12	13	14	15	16	17	18	19	20
9	9	10	11	12	13	14	15	16	17	18	19	20	21
10	10	11	12	13	14	15	16	17	18	19	20	21	22
11	11	12	13	14	15	16	17	18	19	20	21	22	23
12	12	13	14	15	16	17	18	19	20	21	22	23	24

REPRODUCIBLE #6

Multiplication Table

1	2	3	4	5	6	7	8	9	10	11	12
2	4	6	8	10	12	14	16	18	20	22	24
3	6	9	12	15	18	21	24	27	30	33	36
4	8	12	16	20	24	28	32	36	40	44	48
5	10	15	20	25	30	35	40	45	50	55	60
6	12	18	24	30	36	42	48	54	60	66	72
7	14	21	28	35	42	49	56	63	70	77	84
8	16	24	32	40	48	56	64	72	80	88	96
9	18	27	36	45	54	63	72	81	90	99	108
10	20	30	40	50	60	70	80	90	100	110	120
11	22	33	44	55	66	77	88	99	110	121	132
12	24	36	48	60	72	84	96	108	120	132	144

Subtraction Action

Name:_____

Write a number (less than 100) in each of the four squares: A, B, C, and D. Begin subtracting and write the difference between A & B, B & C, C & D, and D & A in the large circles shown by the arrows. Continue by subtracting the numbers in the circles, then write that difference in the smaller squares. Continue this way to find a pattern.

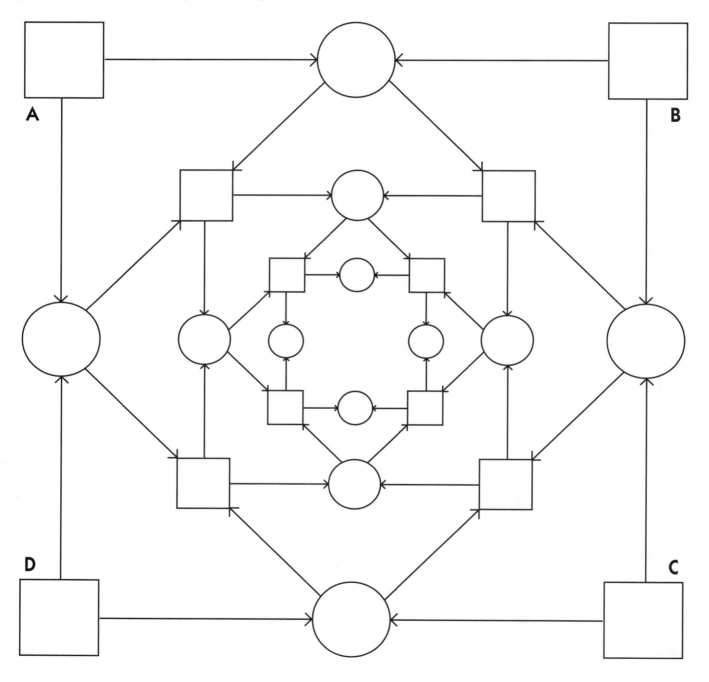

Guesstimate / Bestimate

This Week's Estimation Question:_____

Student/Group	Guesstimate	Bestimate

Putting Statistics on the Line

Today's Topic:

_____

Sample size: _____   Number of values being evaluated

Range: _____   Difference between highest and lowest values

Mode: _____   The value(s) that appears most frequently in the set of data

Mean: _____   The arithmetic average (add all values, then divide by the sample size)

Median: _____   The middle value, found after values are placed in order

REPRODUCIBLE #10

INDEX

Note: Page numbers in italics refer to reproducibles to be used with activities.